"With every new technology, we overestimate how quickly people change their behavior. This dot-com cult classic compares Web fever to the awe of the telegraph."—*Wall Street Journal*

"A fascinating walk through a pivotal period in human history."—*USA Today*

"A new technology will connect everyone! It's making investors rich! It's the Internet boom—except Samuel Morse is there!"—*Fortune*

"[The telegraph's] capacity to convey large amounts of information over vast distances with unprecedented dispatch was an irresistible force, causing what can only be called a global revolution."—*Washington Post*

"Richly detailed and immensely entertaining . . . Standage's writing is colorful, smooth and wonderfully engaging . . . a delightful book."—*Smithsonian Magazine*

"One of the most fascinating books of the dot-com era." —*Financial Times*

"An entertaining primer on a complex subject of increasing interest."—*Los Angeles Times Sunday Book Review*

"Standage tells his fascinating story in an engaging, readable style, from the moment a bunch of Carthusian

monks get suckered into a hilarious human electrical-conductivity experiment in 1746 to the telegraph's eventual eclipse by the telephone. If you've ever hankered for a perspective on media Net hype, this book is for you."
—*Wired*

"Standage has written a lively book on the telegraph and its roles in helping 19th century business and technology grow . . . *The Victorian Internet* demonstrates engagingly that not even 21st century technology is totally new."
—*Denver Post*

"This book should be essential reading for those caught up in our own information revolution."—*Christian Science Monitor*

"Standage's story is rich with anecdotes, bustling with a cast of idealists and eccentrics."—*BookPage*

"An admirably efficient and concise telling of the story of the rise and decline of the telegraph. As with all good case histories, this one excites the mind with parallels to present-day experience."—Henry Petroski, author of *The Pencil: A History of Design and Circumstance*

"[*The Victorian Internet*] is well worth reading, not only for the fascinating story it offers of early successes in global communication but also for the personal stories it relates.

An extraordinary book!"—Vinton Cerf, co-inventor of the Internet

"An inspired and utterly topical rediscovery of the emergence of the earliest modern communications technology."—William Gibson, author of *All Tomorrow's Parties*

"A lively, short history of the development and rapid growth a century and a half ago of the first electronic network, the telegraph, Standage's book debut is also a cautionary tale in how new technologies inspire unrealistic hopes for universal understanding and peace, and then are themselves blamed when those hopes are disappointed."
—*Publishers Weekly*

"A fascinating overview of a once world-shaking invention and its impact on society. Recommended to fans of scientific history."—*Kirkus Reviews*

"This lively, anecdote-filled history reveals that the telegraph changed the world forever—from a hand-carried-message world to an instantaneous one . . . Standage has it all here, including the role the telegraph played in war (Crimea), spying (the Dreyfus affair, in which Captain Dreyfus was first betrayed and then saved by a telegram), and even love (sort of the first chat rooms, to use an Internet term)."
—*Booklist*

THE VICTORIAN INTERNET

THE VICTORIAN

BLOOMSBURY

NEW YORK · LONDON · OXFORD · NEW DELHI · SYDNEY

InTerneT

The Remarkable
Story of the
Telegraph and
the Nineteenth
Century's
On-line Pioneers

TOM
STanDaGe

2Bloomsbury USA
An imprint of Bloomsbury Publishing Plc

1385 Broadway   50 Bedford Square
New York   London
NY 10018   WC1B 3DP
USA   UK

www.bloomsbury.com

BLOOMSBURY and the Diana logo are trademarks of Bloomsbury Publishing Plc

First published in the United States by Walker & Company in 1998
First paperback edition published in 2007
This paperback edition published in 2014

Illustrations on pages 11, 15, 26, 38, 131, 176, 187, and 191 appear courtesy of the Cable & Wireless
Archive, London. Illustrations on pages 36, 80, and 204 used by permission of Warwick Leadlay Gallery,
Greenwich, London. Illustrations on pages 31 and 34 used by permission of the Science and Society Picture
Library, London. Illustrations on pages 60, 103, 142, and 196 used by permission of Culver Pictures.

ISBN: HC: 978-0-8027-1342-4
PB: 978-1-62040-592-5
ebook: 978-1-8027-1879-2

LIBRARY OF CONGRESS CATALOGING-IN-PUBLICATION DATA
Standage, Tom.
The Victorian Internet: the remarkable story of the telegraph and
the nineteenth century's on-line pioneers/Tom Standage.
p. cm.
Includes bibliographical references and index.
ISBN 978-0-8027-1342-4 (hardcover)
1. Telegraph—History.
I. Title.
HE7631.S677 1998
384.1'09—dc21   98-24959

7 9 10 8

Typeset by Coghill Composition Company
Printed and bound in the U.S.A. by Sheridan

To find out more about our authors and books visit www.bloomsbury.com. Here you will find extracts,
author interviews, details of forthcoming events and the option to sign up for our newsletters.

Bloomsbury books may be purchased for business or promotional use.
For information on bulk purchases please contact Macmillan Corporate and
Premium Sales Department at specialmarkets@macmillan.com.

To Dr. K

contents

Foreword to the new edition

I HAVE THREE COPIES of *The Victorian Internet*. The first is an uncorrected advance copy, the second is a hardcover version, and the third is the 2007 paperback edition, with an afterword by author Tom Standage. I found my handwritten note in the uncorrected copy, dated 1998. It reads:

> This brief, intense and insightful account of the development of the telegraph itself kept me fixated from the first to last page. It seems to be thoroughly researched and enjoyably chronicled. The parallels with the Internet illustrate the ease with which our modern-

day hubris makes itself embarrassingly visible. Claims for the virtues of the Internet are echoes of the same claims made for telegraphy, telephony, radio, and television.

One simply must read it to put the Internet into perspective. And, besides, it is filled with truly amusing anecdotes of the main actors and occasional screwballs of the era.

It is now 2013 and fifteen years have elapsed since Tom wrote his first version. Even his 2007 afterword can use some updating after only six years. He is quite correct to observe that wirelessness is reasserting itself as mobile phones, pads, or tablets occupy a place once adorned mostly by desk and laptop computers. Perhaps even more prophetic, we are entering a period in which virtually any appliance, no matter how large or small, may become part of the vast and connected world of the Internet. The phrase "Internet of things" has become popular. At Google, there are self-driving cars that take advantage of wireless connectivity to vast quantities of information: all the cars are learning from each other and from information gathered globally from "crowds" of users contributing geographically indexed information. Google Glass, a wearable device, places computing power and sensory awareness associated with the wearer. The computer is sharing the visual and auditory "sensorium" with us and helping us understand where we are, what is around us, and what might be of interest to us.

Thanks to improvements in microelectronics, battery-powered devices, including a wide range of sensors and actuators, are becoming part of a global artificial fabric of information. Where we once turned to books, other print publications, and even television and radio, we now turn to the Net to search for information, to be alerted to information of interest, to communicate with known and unknown colleagues, to share our experiences, discoveries, and knowledge. We entertain and we are entertained through the fabric of this system. We learn, teach, transact, produce, consume, plan, execute, and collaborate online.

Some of these phenomena might have been predictable as we experienced the effects of rapid delivery of information through the telegraph system. When it was expanded through intercontinental cables and eventually radio, one could try to imagine the effect of the stunningly brief delay for disseminating important information. Our ability to make such predictions is limited in some ways by our ability to imagine the scale and scope of systems of this kind. Time and space are not constraints on the Internet, unlike broadcast radio and television where time is limited or in print publications where space is equally at a premium. The users of the system determine how much time they are willing to spend exploring the vastness of the information that has been and continues to be produced.

There is a dark side that cannot be ignored. Our

dependence on digital technology and digitized information also creates vulnerability. Our devices may ingest digital viruses, worms, Trojan horses, and other malware. Our
daily lives may be disrupted. Our critical infrastructure
damaged or disabled. We have created our own liabilities
to go along with the extraordinary utility of our connected
universe. We are thus reminded that we have challenges to
overcome as privacy is eroded, reliability is undermined,
integrity is compromised, and the system turns our dependence into hazard. There is a story by E. M. Forster titled
"The Machine Stops," written in 1909.* A society depen-
dent on the Machine is suddenly thrown into turmoil as
the sustaining benefits of the Machine abruptly halt.

Tom's excellent work reminds us again and again that
our optimism and enthusiasm must be leavened by
thoughtful perspective. And this his enduring story of the
telegraph provides, at least as I read it.

Vinton Cerf
April 2013

---

*Forster, E. M. "The Machine Stops." *The Oxford and Cambridge
Review*, November 1909.

preface

I n the nineteenth century there were no televisions, airplanes, computers, or spacecraft; nor were there antibiotics, credit cards, microwave ovens, compact discs, or mobile phones.

There was, however, an Internet.

During Queen Victoria's reign, a new communications technology was developed that allowed people to communicate almost instantly across great distances, in effect shrinking the world faster and further than ever before. A worldwide communications network whose cables spanned continents and oceans, it revolutionized business practice, gave rise to new forms of crime, and inundated its users with a deluge of information. Romances blossomed over the wires. Secret codes were devised by some users and cracked by others. The benefits of the network were relentlessly hyped by its advocates and dismissed by the skeptics. Governments and regulators tried and failed to control the new medium. Attitudes toward everything from news gathering to diplomacy had to be completely rethought. Meanwhile, out on the wires, a technological subculture with its own customs and vocabulary was establishing itself.

Does all this sound familiar?

Today the Internet is often described as an information superhighway; its nineteenth-century precursor, the electric telegraph, was dubbed the "highway of thought." Modern computers exchange bits and bytes along network cables; telegraph messages were spelled out in the dots and dashes of Morse code and sent along wires by human operators. The equipment may have been different, but the telegraph's impact on the lives of its users was strikingly similar.

The telegraph unleashed the greatest revolution in communications since the development of the printing press. Modern Internet users are in many ways the heirs of the telegraphic tradition, which means that today we are in a unique position to understand the telegraph. And the telegraph, in turn, can give us a fascinating perspective on the challenges, opportunities, and pitfalls of the Internet.

The rise and fall of the telegraph is a tale of scientific

discovery, technological cunning, personal rivalry, and cutthroat competition. It is also a parable about how we react to new technologies: For some people, they tap a deep vein of optimism, while others find in them new ways to commit crime, initiate romance, or make a fast buck—age-old human tendencies that are all too often blamed on the technologies themselves.

This is the story of the oddballs, eccentrics, and visionaries who were the earliest pioneers of the on-line frontier, and the global network they constructed—a network that was, in effect, the Victorian Internet.

THE VICTORIAN INTERNET

1.

THe MOTHer
OF aLL neTWOrKS

telegraph, *n.*—a system of or instrument for
sending messages or information to a distant place;
v.—to signal (from French *TÉLÉGRAPHE*)

On an APRIL DAY in 1746 at the grand con-
vent of the Carthusians in Paris, about two
hundred monks arranged themselves in a long, snaking
line. Each monk held one end of a twenty-five-foot iron
wire in each hand, connecting him to his neighbor on either side. Together, the monks and their connecting wires
formed a line over a mile long.

Once the line was complete, the abbé Jean-Antoine
Nollet, a noted French scientist, took a primitive electrical
battery and, without warning, connected it to the line of
monks—giving all of them a powerful electric shock.

Nollet did not go around zapping monks with static

electricity for fun; his experiment had a serious scientific objective. Like many scientists of the time, he was measuring the properties of electricity to find out how far it could be transmitted along wires and how fast it traveled. The simultaneous exclamations and contortions of a mile-long line of monks revealed that electricity could be transmitted over a great distance; and as far as Nollet could tell, it covered that distance instantly.

This was a big deal.

It suggested that in theory, it ought to be possible to harness electricity to build a signaling device capable of sending messages over great distances incomparably faster than a human messenger could carry them.

At the time, sending a message to someone a hundred miles away took the best part of a day—the time it took a messenger traveling on horseback to cover the distance. This unavoidable delay had remained constant for thousands of years; it was as much a fact of life for George Washington as it was for Henry VIII, Charlemagne, and Julius Caesar.

As a result, the pace of life was slow. Rulers dispatched armies to distant lands and waited months for news of victory or defeat; ships sailed over the horizon on epic voyages, and those on board were not seen or heard from again for years. News of an event spread outward in a slowly growing circle, like a ripple in a pond, whose edge moved no faster than a galloping horse or a swift-sailing ship.

To transmit information any more quickly, something that moved faster than a horse or a ship was clearly required. Sound, which travels at a speed of about twelve miles per minute, is one means of speedier communication. If a church bell strikes one o'clock, a monk standing in a field half a mile away knows what time it is about two seconds later. A horse-borne messenger, in contrast, setting out from the church precisely on the hour to deliver the message "It is one o'clock," would take a couple of minutes to cover the same distance.

Light also offers an expeditious way to communicate. If the monk has keen eyesight and the air is clear, he may be able to make out the hands of the church clock. And since light (which travels at nearly 200,000 miles per second) covers short distances almost instantly, the information that it is a particular time of day effectively travels from the clock face to the monk in what seems to be no time at all.

Now experiments by Nollet and others showed that electricity also seemed capable of traveling great distances instantaneously. Unlike light, electricity could be transmitted along wires and around corners; a line of sight from one place to another was not needed. This meant that if an electric shock was administered at one o'clock via a half-mile-long wire running from the church to a distant monk, he would know exactly what time it was, even if he was underground or indoors or otherwise out of sight of the clock tower. Electricity held out the promise of high-speed signaling from one place to another, at any time of day.

But the advantage of a horse-borne message was that it could say anything at all; instead of saying "It is one o'clock," it could just as easily say "Come to lunch" or "Happy birthday." An electrical pulse, on the other hand, was like the strike of a church bell, the simplest of all possible signals. What was needed was a way to transmit a complicated message using simple signals. But how could it be done?

SINCE THE LATE sixteenth century there had been persistent rumors across Europe of a magical device that allowed people many miles apart to spell out messages to each other letter by letter. There was no truth in these tales, but by Nollet's time the stories had acquired the status of what might today be called an urban myth. Nobody had actually seen one of these devices, which relied on magical "sympathetic" needles that could somehow influence each other over great distances, but they were widely believed to exist. Cardinal Richelieu, for example, the ruthless and widely feared first minister of France, was thought to have a set because he always seemed so well informed about goings-on in distant places. (Then again, he was also thought to be the owner of a magical all-seeing eye.)

Perhaps the best-known description of the sympathetic needles was published by Famianus Strada, a learned Italian who provided a detailed explanation in his book *Prolusiones Academicae*, published in 1617. He wrote of "a species of lodestone which possesses such virtue, that if two needles be touched with it, and then balanced on separate pivots, and the one turned in a particular direction, the other will sympathetically move parallel to it." Each needle, he explained, was to be mounted in the center of a dial, with the letters of the alphabet written around its edge. Turning one of the needles to point to the letter "A" on its dial would then supposedly cause the other sympathetic needle to indicate the same letter. And this was all said to work no matter how far apart the two needles were. By indicating several letters in succession, a message could then be sent from one place to another.

"Hither and thither turn the style and touch the letters, now this one, and now that," wrote Strada. "Wonderful to relate, the far-distant friend sees the voluble iron tremble without the touch of any person, and run now hither, now thither: he bends over it, and marks the teaching of the rod. When he sees the rod stand still, he, in his turn, if he thinks there is anything to be answered, in like manner, by touching the various letters, writes it back to his friend."

The story of the needles was based on a germ of truth: There are indeed naturally occurring minerals, known as lodestones, which can be used to magnetize needles and other metallic objects. And if two magnets are placed on pivots very close together, moving one will indeed cause the other to move in response, as a result of the interaction of their magnetic fields. But it is not the case that the two magnets will always remain parallel, and the effect is only noticeable when they are right next to each other. The kind of needles described by Strada, which could interact over great distances, simply did not exist.

But that didn't stop people from talking about them. One wily salesman is even said to have tried to sell a set of needles to Galileo Galilei, the Italian astronomer and physicist. A firm and early believer in experimental evidence and direct observation, Galileo demanded a demonstration of the needles on the spot. The salesman refused, claiming that they worked only over very great distances. Galileo laughed him out of town.

Yet the talk of magic needles continued, along with the research into the properties of electricity. But no progress toward a practical signaling device was made until 1790. When the breakthrough finally came, it didn't involve needles or lodestones or electric wires; in fact, it was surprising that nobody had thought of it sooner.

CLOCKS AND COOKING PANS hardly seem the stuff of which communications revolutions are made. But that was what Claude Chappe ended up using for his first working signaling system.

Chappe was one of many researchers who had tried

and failed to harness electricity for the purpose of sending messages from one place to another. Born into a well-to-do French family, he planned on a career as a member of the clergy but was derailed by the French Revolution in 1789. He took up scientific research instead, concentrating on physics and, in particular, the problems associated with building an electrical signaling system. Having made no more progress than anyone else, he decided to try a simpler approach. Before long he had figured out a way to send messages using the deafening "clang" made by striking a casserole dish—a sound that could be heard a quarter of a mile away—in conjunction with two specially modified clocks. They had no hour or minute hands, just a second hand that went twice as fast as usual, completing two revolutions per minute, and a clock face with ten instead of the usual twelve numbers around its edge.

To send a message, Claude Chappe and his brother René, stationed a few hundred yards apart behind their parents' house, would begin by synchronizing their clocks. Claude would make a "clang" as the second hand on his clock reached the twelve o'clock position, so that his brother could synchronize his clock accordingly. Claude could then transmit numbers by going "clang" as the second hand passed over the number on the clock face that he wished to send. Using a numbered dictionary as a codebook, the Chappe brothers translated these numbers into letters, words, and phrases, and thus sent simple messages. It is uncertain how their original code worked, but the brothers probably transmitted digits in twos or three and looked up the resulting two- or three-digit number in the codebook to see what word or phrase it corresponded to.

In other words, a complicated message could be sent using simple signals. However, the problem with this design (apart from the incessant clanging noise) was that the person receiving the message had to be within earshot of the sender and, depending on the direction of the wind, this was a few hundred yards away at most. Rather than replacing the copper pans with something louder, Chappe realized that it would be simpler to replace the audible signal with a visible one.

So out went the casserole dishes, and in their place was substituted a pivoting wooden panel, five feet tall, painted black on one side and white on the other. By flipping it from one color to the other as the second hand passed over a particular number, Chappe could transmit that number. This improved design had the obvious advantage that it allowed messages to be sent over very great distances very quickly—particularly if a telescope was used to observe the panel.

At 11 A.M. on March 2, 1791, Chappe and his brother used their black-and-white panels, clocks, telescopes, and codebooks to send a message between a castle in their hometown of Brûlon, in northern France, and a house in Parcé, ten miles away. In the presence of local officials, it took them about four minutes to transmit a phrase chosen by the local doctor—"SI VOUS RÉUSSISSEZ, VOUS SEREZ BIEN-TÔT COUVERT DE GLOIRE" (IF YOU SUCCEED, YOU WILL SOON BASK IN GLORY)—from one location to the other.

Chappe wanted to call his invention the *tachygraphe*—from the Greek for "fast writer"—to signify the unprecedented speed with which it transmitted information. However, he was talked out of it by his friend Miot de Mélito, a government official and classical scholar, who suggested the name *télégraphe*, or "far writer," instead.

Et voilà: The telegraph was born.

Having shown that his invention worked, Chappe started to galvanize support for it in Paris with the help of another of his brothers, Ignace, who had been elected to the ruling Legislative Assembly. But the turmoil of revolutionary France was a difficult environment in which to start promoting a new invention, and Ignace didn't get very far. When the Chappe brothers staged another test in the town of Belleville, near Paris, in 1792, their apparatus was destroyed by a mob who suspected that they were trying to communicate with royalist prisoners being held in Temple Prison. The Chappes were lucky to escape with their lives.

By this time Claude Chappe had found a way to do without the synchronized clocks: He devised a completely new design that consisted of two small rotating arms on the end of a longer rotating bar. This bar, called the regulator, could be aligned horizontally or vertically, and each of the small arms, called indicators, could be rotated into one of seven positions in forty-five-degree increments. The design allowed for a total of 98 different combinations, 6 of which were reserved for "special use," leaving 92 codes to represent numbers, letters, and common syllables. A special codebook with 92 numbered pages, each of which listed 92 numbered meanings, meant that an additional 92 times 92, or 8,464, words and phrases could be represented by transmitting two codes in succession. The first indicated the page number in the codebook, and the second indicated the intended word or phrase on that page.

Abraham-Louis Bréguet, the noted clock maker, built Chappe a clever control mechanism for his new design: Through a system of pulleys, a scaled-down model of the rotating arms could be used to control the positions of a much bigger set of arms. The big arms could then be mounted on the roof of a tower and controlled from the inside by an operator. Chappe believed it would be possible to send messages quickly over great distances by constructing several such towers a few miles apart in a long line, each within sight of the next.

In 1793, Chappe sent details of this new design to the National Convention, the ruling body that had replaced the Legislative Assembly. His proposal was picked up by Charles-Gilbert Romme, president of the Committee of Public Instruction, who grasped its potential and suggested that the convention fund an experiment to evaluate its military applications.

Chappe-style optical telegraph, showing arm positions corresponding to different letters. Mounted on the roof of a tower, the arms were controlled from the inside by an operator.

A committee was duly appointed, consisting of Joseph Lakanal, a respected scientist; Louis Arbogast, a professor of mathematics; and Pierre Claude François Daunou, a legislator and historian. Money was allocated for the construction of a line of three telegraph stations in Belleville, Écouen, and Saint-Martin-du-Tertre, spanning a distance of about twenty miles. If a message could be passed along a network of three towers successfully, the system would obviously work with a larger number of towers over greater distances. Following the Chappe brothers' run-in with the Paris mob, the mayors of each of the three towns were made responsible for the safety of the telegraphs and their operators.

Within a few weeks the towers had been constructed, and the committee was invited to a demonstration on July 12, 1793. Transmission of the first message began at 4:26 P.M., with two operators at each station, one operating the little whirling arms, the other watching the next station through a telescope. The position of the arms at the sending station was reported by the observer in the middle station, where the operator would then move the arms to form the same signal; each signal was held in place for a few seconds, and the message rippled down the line to the receiving station. The three telegraph towers took eleven minutes to send a rather boring message ("DAUNOU HAS ARRIVED HERE. HE ANNOUNCED THAT THE NATIONAL CONVENTION HAS JUST AUTHORIZED HIS COMMITTEE OF GENERAL SECURITY TO PUT SEALS ON THE PAPERS OF THE DEPUTIES") along the line one way, and nine minutes to send an equally nondescript reply back again. But the experiment was a success; the committee members, and Lakanal in particular, were highly impressed.

Two weeks later Lakanal addressed the convention in glowing terms on the potential of this great new invention, and how wonderful it was that it had been invented by a Frenchman. "What brilliant destiny do science and the arts not reserve for a republic which, by the genius of its inhabitants, is called to become the nation to instruct Europe," he gushed. Much of his enthusiasm seems to have stemmed from the potential application of the telegraph in holding the newly founded French Republic together, by ensuring that the central government in Paris could keep a firm grip on the provinces. At any rate, following his speech, the construction of a fifteen-station line from Paris to Lille, about 130 miles to the north, was proposed. Chappe was put on a government salary, complete with the use of a horse.

The Paris-Lille line, the first arm of the French State Telegraph, started operation in May 1794, and on August 15 it was used to report the recapture of a town from the Austrians and Prussians within an hour of the battle's end. As the French army advanced north into Holland, further victories were reported via the telegraph, and the government's appreciation for Chappe's invention grew. By 1798, a second line had been built to the east of Paris as far as Strasbourg, and the Lille line had been extended to Dunkirk.

Napoleon Bonaparte, who seized power in 1799, was a firm believer in the telegraph; he ordered further extension of the network, including the construction of a line to Boulogne in preparation for an invasion of England. He also asked Abraham Chappe, Claude's younger brother, to design a telegraph capable of signaling across the English Channel. (A successful prototype was built and tested between Belleville and Saint-Martin-du-Tertre, the two stations on the original experimental line, the distance between which was roughly equivalent to the shortest distance across the Channel. The station on the French side was later installed in Boulogne, but Napoleon's plans for the invasion never materialized, so neither did the British station.) In 1804, Napoleon ordered the construction of a line from Paris to Milan, via Dijon, Lyons, and Turin. This was to expand the network farther than ever before.

Lakanal's prediction had by this time come true, and France had indeed become "the nation to instruct Europe." Recognizing the military value of the telegraph, the governments of other European countries, notably Sweden and Britain, had quickly copied Chappe's design or adopted variations upon it. In Britain, the Admiralty ordered the construction of a line of telegraph towers in 1795 to facilitate communication between London and the ports of the south coast during the war with France. The British telegraph was designed by George Murray, a clergyman and amateur scientist, and it consisted of six wooden shutters, each of which could be opened or closed to give sixty-four possible combinations (since $64 = 2 \times 2 \times 2 \times 2 \times 2 \times 2 = 2^6$). Soon, telegraph towers were springing up all over Europe.

THE TELEGRAPH SYSTEM was rightly regarded as a technological wonder of its time. An entry in the 1797 edition of the *Encyclopaedia Britannica* strikes a note of technological optimism that sounds familiar today: "The capitals of distant nations might be united by chains of posts, and the settling of those disputes which at present take up months or years

British shutter telegraph, 1797. Each of the six panels could be open (horizontal) or closed (vertical, as shown), giving a total of sixty-four different combinations.

might then be accomplished in as many hours." The author of the encyclopedia entry also suggested that it might be worth opening up the network to paying customers: "An establishment of telegraphs might then be made like that of the post; and instead of being an expense, it would produce a revenue."

Chappe also had all sorts of ambitious plans for his invention; he hadn't intended its use to be so predominantly military in nature, and wanted to promote its employment in business. He suggested a European network relaying commodity prices between Paris and Amsterdam, Cádiz, and even London via a cross-Channel telegraph. He also advocated a state-sanctioned daily national news bulletin. But Napoleon rejected both ideas, though he did agree to allow the weekly transmission of winning national lottery numbers. This meant the numbers were known throughout the country on the day of the draw, dramatically reducing the level of cheating.

But despite the success of his invention, Claude Chappe was not a happy man. He faced increasing criticism from rival inventors, who claimed either to have invented superior forms of telegraph or to have had the idea for the telegraph before him. Even his former associate, the clock maker Bréguet, turned against him and claimed to have contributed far more to Chappe's design than simply the control mechanism. Chappe sank into a deep depression and became increasingly paranoid, even going as far as to accuse his rivals of having slipped something into his food when he suffered from a bout of food poisoning. Finally, on January 23, 1805, he killed himself by jumping into the well outside the Telegraph Administration building in Paris. He was buried under a tombstone decorated with a telegraph tower showing the sign for "at rest."

Even so, his invention continued to flourish; by the mid-1830s, lines of telegraph towers stretched across much of western Europe, forming a sort of mechanical Internet of whirling arms and blinking shutters, and passing news and official messages from one place to another. The continental network eventually reached from Paris to Perpignan and Toulon in the south, Amsterdam in the north, and from Brest in the west to Venice in the east, with other networks in Finland, Denmark, Sweden, Russia, and Britain bringing the total number of telegraph towers in Europe to almost a thousand.

a S THE neTWOrk Grew, telegraph fever took hold in Britain, where amateur scientists, part-time inventors, and quacks were soon hard at work devising improvements to the nation's telegraphs. The Admiralty, which had spent much of the eighteenth century fending off idiotic suggestions about the best way to determine longitude at sea, now found itself on the receiving end of dozens of equally well meant but often crackpot schemes suggesting ways in which the telegraphs could be made faster or cheaper, or both. Some inventors advocated modifications to the six-panel shutter design that had been adopted in Britain; others proposed new and improved codebooks; and some called for the whole system to be scrapped and replaced with a completely new approach. One or two correspondents even claimed to have constructed telegraphs that used electricity.

The earliest suggestion of a scheme for using electricity to send messages had appeared in the *Scots' Magazine* of February 17, 1753. A letter from an unknown author, who signed himself simply "C. M.," was published under the heading "An Expeditious Method of Conveying Intelligence." The letter described a simple signaling system consisting of a wire for each letter of the alphabet, and a frictional generator that sent shocks down the wires. However, there is no evidence that C. M. ever actually built such a telegraph, and his identity remains a mystery.

But between the publication of C. M.'s letter and Queen Victoria's accession in 1837, at least sixty experimental electric telegraphs based on various electric and electrochemical processes are known to have been constructed by a number of researchers. Different designs used bubbling chemicals, sparks, or the twitching of pith balls to detect tiny electric shocks sent along wires. Some, like the telegraph described by C. M., involved twenty-six wires (one for each letter of the alphabet), while others used combinations of a smaller number of wires. But the scientists who constructed them worked independently, and each had to start from scratch—and, crucially, none of them ever managed to stage a decisive demonstration like the one Chappe had used to prove the value of his optical system.

In fact, as far as most people were concerned, so little progress appeared to have been made toward the goal of a practical electric telegraph compared to the highly successful optical design that anyone who expressed an interest in electric telegraphy was regarded as something of an eccentric. As one satirical verse of 1813 put it:

> Our telegraphs, just as they are, let us keep,
> They forward good news from afar;
> And still may send better—that Boney's asleep

And ended oppression and war.
Electrical telegraphs all must deplore,
Their service would merely be mocking;
Unfit to afford us intelligence more
Than such as would really be shocking.

One example of a working electric telegraph was constructed in 1816 by a twenty-eight-year-old Englishman,
Francis Ronalds. Similar to Chappe's original design, it
involved synchronized clocks whose dials were marked
with letters, rather like the dials supposedly used with the
sympathetic needles. Instead of hands, each clock had a
rotating disk with a notch cut into it so that only one of
the letters on the dial was visible at any time. And instead
of the clash of copper pans or the turning of a black-and-
white shutter to signal each letter, Ronalds used electricity.
When an electric shock, generated by a frictional generator, was sent down a wire from the transmitting station, it
caused a pair of pith balls suspended from the wire at the
receiving station to become electrically charged; the balls
twitched as they briefly repelled one another, and the operator would write down the letter indicated on the dial at
that moment.

Ronalds set up an experimental system in his garden
and wrote to the government, requesting an interview with
Lord Melville, the first lord of the Admiralty, to arrange a
demonstration. "Why has no serious trial yet been made
of the qualifications of so diligent a courier?" he asked.

"And if he should be proved competent to the task, why should not our kings hold councils at Brighton with their ministers in London? Why should not our government govern at Portsmouth almost as promptly as at Downing Street? Let us have electric conversazione offices, communicating with each other all over the kingdom, if we can."

However, along with all the other suggestions for ways to improve the telegraphs, Ronalds's farsighted ideas were politely but firmly rejected. John Barrow, secretary of the Admiralty, wrote back saying that since the war with France had ended, the telegraph system was in no need of improvement. "Telegraphs of any kind are now wholly unnecessary," he wrote, adding that "no other than the one now in use will be adopted."

The Admiralty's position was understandable; it could hardly waste its time investigating every quack's claim to have done the impossible and built a practical electric telegraph. Ronalds never got the chance to demonstrate his invention, but, surprisingly, he seems to have taken his rejection very well. "Everyone knows," he noted, "that telegraphs have long been great bores at the Admiralty." He gave up telegraphy and took up weather forecasting instead.

ULTIMATELY, the success of the optical telegraph designs inspired by Chappe was limited because they were so expensive to run. They required shifts of skilled operators at each station and involved building towers all over the place, so that only governments could afford to run them; and their limited information-carrying capacity meant they were just used for official business. Optical telegraphs had shown that complex messages could be sent using combinations of simple signs; but other than noticing the appearance of a tower on top of the nearest hill, most people's lives were not directly affected. (Today, all that is left of the original telegraph network is a few place-names; several hills are still known as Telegraph Hill.)

As well as being expensive, optical telegraphs suffered from the drawback of not working in the dark, despite various experimental schemes that involved the use of colored lanterns on the end of the indicator arms. But at least the fall of darkness could be predicted; fog and mist, on the other hand, could arise at any time. When choosing the locations for new telegraph towers, one had to ensure that there were no marshes, rivers, or lakes along the line of sight between adjacent towers; local inhabitants were often consulted to determine the likelihood of mists arising.

If it could ever be constructed, however, a practical electric telegraph would work over any terrain, in any kind of weather, at any time of the day or night. It would be able to send messages around corners and over mountains. Yet for all its supposed advantages, and despite the work of Ronalds and others, the electric telegraph was still widely believed to be nothing more than an impossible dream.

2.

STRANGE, FIERCE FIRE

But one morning he made him a slender wire,
As an artist's vision took life and form,
While he drew from heaven the strange, fierce fire
That reddens the edge of the midnight storm;
And he carried it over the Mountain's crest,
And dropped it into the Ocean's breast;
And Science proclaimed, from shore to shore,
That Time and Space ruled man no more.
    —from "The Victory," a poem written
    in tribute to Samuel Morse, 1872

TODAY, even a CHILD could build an electric telegraph. All you need is a battery, a bulb, and some wire to connect the two. I hold on to the battery, while you sit, some distance away, by the bulb; I run connecting wires from one to the other; and by making and breaking the circuit at my end, I can get the light to flash on and off at your end. Provided we have agreed upon a suitable way to represent letters of the alphabet, I can send you messages. (One obvious but rather inefficient scheme would be for one flash to signify "A," two to signify "B," and so on.)

In the early nineteenth century, of course, batteries and bulbs weren't readily available. Crude batteries like the one Nollet used to shock the monks had given way to the voltaic cell, invented around 1800 by Alessandro Volta, which works on the same principle as a modern battery. Rather than simply discharging to give one brief jolt of current, voltaic cells could drive current in an orderly fashion around an electric circuit.

But it was another eighty years before the American inventor Thomas Edison would invent the lightbulb, so there was still no easy way to detect the presence of electricity in a wire. Experimenters used electricity to cause pith balls to twitch, to trigger chemical reactions, and cause sparks. But experimental telegraphs based on such cumbersome means of detecting current (such as the one built by Ronalds) were unreliable and unwieldy, and never got very far.

The breakthrough came in 1820 when Hans Christian Oersted, a Danish physicist, observed that electric current flowing in a wire gives rise to a magnetic field, a phenomenon known as electromagnetism. This magnetic field can then be detected through its effect on another object: As Oersted discovered, it will cause a nearby compass needle to move. For the first time, there was a reliable, repeatable, and practical way to detect electricity. (Ironically, it depended on magnetism—the principle that had been the basis of the myth of the sympathetic needles.)

Two new inventions quickly followed: the galvanometer, which indicates the flow of current by the deflection of a rotating needle, and the electromagnet, a coil of wire that behaves just like a permanent magnet—but only as long as current is flowing through it. Together with the new voltaic battery, either could be used as the basis of an electric telegraph.

But those who tried to build telegraphs based on electromagnetic principles soon ran into a new problem. Even when equipped with the latest batteries and electromagnets, some people seemed to have less success than others when they tried signaling over long wires; and nobody could understand why.

In 1824, for example, the British mathematician and physicist Peter Barlow considered the possibility of building an electric telegraph that would send messages using an electromagnet that made a clicking sound as it was switched on and off. "There is only one question which would render the result doubtful: is there any diminution of the effect [of electricity] by lengthening the conducting wire?" he asked. "I found such a diminution with only two hundred feet of wire, as at once to convince me of the impracticability of the scheme."

Barlow was not alone. In their own experiments, many

other scientists had found that the longer the wire they used, the weaker the effects of the electricity at the other end. To those working in the field, a practical electric telegraph seemed as far away as ever.

S a m u e l F. B. m o r s e was born in Charlestown, Massachusetts, in 1791, the year of Chappe's first demonstration of an optical telegraph. He was a johnny-come-lately to the field of electric telegraphy. Had he started building an electric telegraph a little earlier, he might have got home in time for his wife's funeral.

Morse's wife, Lucretia, died suddenly at their home in New Haven, Connecticut, on the afternoon of February 7, 1825, while her husband was away. He was starting to make progress in his chosen career as a painter and had gone to Washington to try to break into the lucrative society portrait business. He had just been commissioned to paint a full-length portrait of the marquis de Lafayette, a military hero, and his career finally seemed to be taking off. "I long to hear from you," he wrote in a letter to his wife on February 10, unaware that she was already dead.

Washington was four days' travel from New Haven, so Morse received the letter from his father telling him of Lucretia's death on February 11, the day before her funeral. Traveling as fast as he could, he arrived home the following week. His wife was already buried. In the United

Samuel F. B. Morse, one of the inventors of the electric telegraph.

States in 1825, messages could still only be conveyed as fast as a messenger could carry them.

Morse was forty-one when he caught the telegraph bug following a chance meeting on board a ship in the mid-Atlantic. In 1832, he was returning to the United States from Europe, where he had spent three years in Italy, Switzerland, and France improving his painting skills and working on a rather harebrained scheme to bring the treasures of the Louvre in Paris to an American audience. On a six-by-nine-foot canvas, he was painting miniature copies of thirty-eight of the Louvre's finest paintings, which he collectively dubbed the *Gallery of the Louvre*. The painting, still unfinished, accompanied Morse onto the sailing packet *Sully*, a fast ship that was carrying mail, together with a small number of well-to-do passengers, across the Atlantic.

His intention was to finish the *Gallery of the Louvre* when he got back to the United States, and then exhibit it and charge admission. It was a scheme typical of Morse: Since 1823, for example, he had been experimenting with a marble-cutting device that would supposedly make copies of any sculpture, with a view to reproducing well-known works of art in large quantities for sale to the public. And as a young man, he had dabbled with various other inventions, including a new kind of water pump, devised in 1817, which he sold to a local fire brigade. But none of his schemes, which typically combined artistic ingenuity with public-spiritedness, had ever been successful; the hapless Morse seems to have stumbled from one money-making idea to another as the mood took him.

As the *Sully* made its way across the ocean, the passengers on board got to know each other quite well, and, two weeks into the voyage, a philosophical discussion at the dinner table one afternoon turned to the matter of electromagnetism. Dr. Charles Jackson of Boston, one of the passengers, knew a good deal about the subject and even had an electromagnet and some other electrical bits and bobs with him on board the ship. In the midst of an explanation, one of the passengers asked Jackson the very question that Nollet's experiment had been trying to answer: How fast did electricity travel along a wire, and how far could it go?

As the electrified monks could have testified back in 1746, and as Dr. Jackson explained, electricity was believed to pass through a circuit of any length instantaneously. Morse was thunderstruck. "If the presence of electricity can be made visible in any desired part of the circuit," he is reputed to have said, "I see no reason why intelligence might not be instantaneously transmitted by electricity to any distance." This, of course, was exactly the reason that so many scientists had spent the best part of a century trying to harness electricity as a means of signaling, but Morse didn't know that. He left the table, went up on deck, and started scribbling in his notebook. Convinced that he was the first to have had the idea, he instantly became obsessed with a new scheme: building an electric telegraph.

Perhaps fortunately, Morse was unaware that other would-be telegraphers had failed after being unable to get signals to travel over long wires. Assuming that the electric side of things would be fairly straightforward, he started thinking about the other half of the problem: a signaling code.

The arms or shutters of an optical telegraph can be arranged in a large number of different combinations, but an electric current can only be on or off. How could it be used to transmit an arbitrary message? As he paced the deck of the *Sully*, Morse swiftly rejected the approach of using a separate electrical circuit for each letter of the alphabet. Next, he considered the possibility of using the clicking of an electromagnet to send numbers in the same way as a church bell, which indicates the hour by the number of chimes. But with this system, it would take nine times as long to send a 9 (9 clicks) as it would to send a 1 (1 click).

Before long, Morse had the idea of using short and long bursts of current—a "bi-signal" scheme that later evolved into the dots and dashes of what we now know as Morse code. He decided upon a series of short and long bursts corresponding to each of the digits from 0 to 9, and sketched them in his notebook. Sending a series of digits, he decided, could then be used to indicate a word in a numbered codebook.

Next, Morse turned to the matter of creating a permanent record of an electric signal so that it could be translated from dots and dashes back into the original message. Together with Jackson, he sketched out a way to record incoming signals on paper automatically, by marking a paper tape with a moving pencil controlled by an electromagnet.

After six weeks at sea, Morse arrived in New York a changed man. He met his brothers Richard and Sidney on the dock and started telling them about his new scheme almost immediately. "Hardly had the usual greetings passed between us three brothers, and while on our way to my house, before he informed us that he had made, during his voyage, an important invention, which had occupied almost all his attention on shipboard," Richard recalled.

Sidney remembered that his brother was "full of the subject of the telegraph during the walk from the ship, and for some days afterward could scarcely speak about anything else." Morse immediately set to work building an electric telegraph.

F OUR YEARS LATER, in 1836, a young Englishman experienced a similar epiphany. William Fothergill Cooke was the son of a professor of anatomy who found himself at loose ends after resigning his commission in the Indian army, and took to making anatomical wax models of dissected cadavers for use in medical training. While studying anatomy in Heidelberg, he happened to attend a lecture about electricity, and before long he too had decided to try his hand at building an electric telegraph.

The lecture Cooke attended included a demonstration of an experimental telegraph system that had been invented by Baron Pavel Lvovitch Schilling, a Russian diplomat, in the mid-1820s. Based on a galvanometer, it used combinations of the left and right swings of the galvanometer needle to indicate letters and numbers. Just as Ronalds had done in Britain, Schilling promoted his invention to his superiors in government, and after many years of lobbying he managed to arrange a demonstration in 1836 in the presence of Czar Nicholas, who was very impressed and gave his approval for the construction of an official

William Fothergill Cooke,
one of the British inventors
of the electric telegraph.

network. But Schilling died shortly afterward, and his telegraphic ambitions died with him.

However, Professor Muncke of Heidelberg University
had a copy of one of Schilling's galvanometers, which he
liked to use to demonstrate the principle of electromagnetism. After attending such a demonstration, Cooke was
"struck with the wonderful power of electricity and
strongly impressed with its applicability to the practical
transmission of telegraphic intelligence." Realizing that
this phenomenon might, as he put it, "be made available
to purposes of higher utility than the illustration of a lecture," Cooke (who had been looking around for a way to
make his fortune) immediately abandoned anatomy and
decided to build an electric telegraph based on an improved version of Schilling's apparatus.

Within three weeks he had built a prototype that combined three of Baron Schilling's needle telegraphs in a single device. It used a system of switches to control three needles via six wires. Each needle could be made to tilt to the left or the right, or could remain unmoved, and different combinations of the three needles' positions signified different letters.

Having built prototypes that were capable of sending messages over wires thirty or forty feet long, Cooke, who had by this time returned to England, was eager to try out his apparatus over greater distances. His friend Burton Lane, a solicitor at Lincoln's Inn in London, gave him the use of his office for three days so that he could lay out a mile of wire. "I had to lay out this enormous length of 1,760 yards in Burton Lane's small office, in such a manner as to prevent one part touching another; the patience required and the fatigue undergone was far from trivial," Cooke wrote in a letter to his family. Worse still, the result was disappointing: His apparatus simply didn't work with longer wires. After a week, he had outstayed his welcome, and Lane wanted his office back.

Meanwhile in New York, Morse, working independently, had come up against exactly the same problem. Although his telegraph worked over short distances, all his experiments with longer wires had failed. Each man realized that there was more to the electric side of building a telegraph than first suspected, and neither had the scientific training to get past this hurdle.

In fact, the problem had already been solved by Joseph Henry, an American physicist, who had managed to get a battery and an electromagnet, connected by 1,060 feet of wire, to ring a bell. In a series of experiments carried out in 1829 and 1830, Henry discovered that getting an electric current to travel through a long wire was all a matter of using the right kind of battery. He found that in conjunction with a suitable electromagnet, a large number of small batteries connected in a row, rather than a single large battery, enabled the signal to travel much farther. But Morse and Cooke, as amateur experimenters, were unaware of Henry's work, even though members of the scientific community on both sides of the Atlantic were familiar with it.

Cooke arranged a meeting with Michael Faraday, the eminent British scientist whose particular area of research at that time was the relationship between electricity and magnetism. Faraday confirmed that Cooke's design for a telegraph was technically sound; but when Cooke enthusiastically offered to explain his design for a perpetual motion machine as well, Faraday, suspecting that he had a quack on his hands, declared that he was pressed for time and showed Cooke the door.

Next, Cooke turned to his friend Peter Roget for advice. Roget is best known today as the compiler of the first thesaurus, but he was also a scientist and had published a treatise on electricity in 1832. He introduced Cooke to Professor Charles Wheatstone, who had made his name

Professor Charles Wheatstone, scientist and co-inventor of the electric telegraph.

through an ingenious series of experiments to determine the velocity of electricity. A meeting was arranged, and Cooke was delighted to discover that Wheatstone had quite a length of wire—four miles of it, in fact—ready for experimentation. He was rather less pleased to hear that Wheatstone had also been carrying out telegraphic experiments of his own. Moreover, since Wheatstone was familiar with Henry's work, he had succeeded in getting signals to travel over long distances where Cooke had failed.

The two men formed an uneasy partnership: Cooke needed Wheatstone's scientific knowledge, so he offered Wheatstone a sixth share in the profits of his device. Wheatstone haughtily proclaimed that he thought it was inappropriate for scientific men to do anything other than publish their results and let others make whatever commercial use of them they wanted; but that if he was to be a partner with Cooke, who was his junior, it would have to be on equal terms. Impressed by what he later described as Cooke's "zeal, ability and perseverance," Wheatstone eventually agreed to a partnership, on the rather childish condition that his name would go first on the documentation.

This sort of behavior was typical of Wheatstone, who was a somewhat prickly character, and whose relationship with Cooke was always highly precarious. By turns painfully shy and incurably arrogant, Wheatstone insisted on referring to the invention of the telegraph in the first person singular and claiming all the scientific credit for himself, as though Cooke were nothing more than a business associate whom he had engaged to promote his invention.

But even though they didn't get on personally, the two men's professional relationship was productive: They had soon devised and patented an improved five-needle telegraph. Each needle could be deflected to the left or right to pick out numbers and letters on a diamond-shaped grid, so there was no need to learn which combination corresponded to which letter. However, the limited number of possible combinations with the five-needle design meant that only twenty letters were included in the telegraphic alphabet; thus "c," "j," "q," "u," "x," and "z" were omitted. Although this design required separate wires between sender and receiver for each needle, it could transmit messages quickly without the need for a codebook.

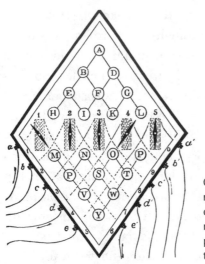

Cooke and Wheatstone's original five-needle electric telegraph. Each needle could be tilted to the left or right, or remain vertical; moving two needles picked out a letter on a diagonal grid (in this case, the letter "v").

 orse had by this time spent five years working on his telegraph, compared to a few months in the case of Cooke and Wheatstone. This was largely because he had got sidetracked into building a vastly overcomplicated design that involved feeding a preprepared rack (or "port rule") of toothed pieces of metal, each representing a letter or number, into the sending apparatus. As the rack passed through the machine, the spacing of the teeth caused long and short pulses of electricity to be transmitted down the wire to the receiver, switching an electromagnet on and off and deflecting a pencil as it drew a line on a moving strip of paper. The long and short pulses were transcribed as a zigzag line, whose wiggles could then be translated from Morse code back into the original message. Morse thought the advantages of this rather convoluted scheme were that messages could be prepared for transmission in advance, and that at the receiving end there would be a permanent record of all incoming messages. It was all rather complicated, and Morse, who was living on a tiny salary after being appointed professor of literature of the arts of design at New York University, frequently had to choose between spending his money on food or components for his telegraph. So it had taken him a long time to build the device.

Having run into the problem of transmitting over long distances, Morse too was guided by a helpful academic. Professor Leonard Gale, who taught chemistry at New York University, was a personal friend of Henry's and suggested changing the battery and improving the receiving electro-magnet. "After substituting the battery of twenty cups for that of a single cup, we sent a message through two hundred feet of conductors, then through 1,000 feet, and then through ten miles of wire arranged on wheels in my own lecture room in the New York University in the presence of friends," Gale recalled. This was the breakthrough that Morse had been seeking.

Morse and Gale teamed up and were soon joined by Alfred Vail, a young man who saw a demonstration of Morse's prototype telegraph and wanted to get involved. In exchange for becoming a partner in the venture, with a share of the patent rights, he agreed to build a complete

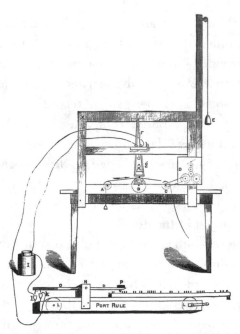

Morse's original telegraph. Winding the handle (L) forced the toothed rack through the transmitting apparatus (P), making and breaking the circuit. At the receiving end, the intermittent current was recorded as a zigzag line on a moving tape (A) by deflecting a pencil (G) with an electromagnet.

set of instruments at his own expense. For the cash-strapped Morse, Vail, who had money, enthusiasm, and practical experience from his father's ironworks, was a godsend.

With Vail on board, Morse's design made progress in leaps and bounds. They did away with the rack and the toothed metal pieces in favor of tapping a key by hand. The zigzag line drawn by the pencil was replaced by an ink pen that rose and fell to inscribe a line of dots and dashes. Morse's number system was also replaced with an alphabetic code, in which each letter was represented by a combination of dots and dashes, thus doing away with the need for numbered codebooks. By counting the number of copies of each letter in a box of printer's type, Morse and Vail designed the code so that the most common letters had the shortest equivalents in code; "E," the most common letter, was represented by a single dot.

As THEY PERFECTED their designs, Morse and Cooke were both aware of the wider significance of their work, although they were still unaware of each other's efforts. Cooke thought the electric telegraph would be useful to governments "in case of disturbances, to transmit their orders to the local authorities and, if necessary, send troops for their support." He also thought it could be used for transmitting stock prices or to help a family in the event of an illness "hastening towards a fatal termination with such rapidity that a final meeting is without the range of ordinary means."

Morse had similar ambitions for his telegraph. One visitor to his rooms recalled that he "believed he had discovered a practical way of using [electromagnetism] as a means of communication and interchange of thought in written language, irrespective of distance and time save that required for manipulation, and that it would ultimately become a daily instrumentality in domestic as well as public life."

Right from the start, Morse was confident that Europe and North America would eventually be connected by a wire that would link telegraph networks on both sides of the Atlantic. He had visions of a wired world, with countries bound together by a global network of interconnected telegraph networks. "If it will go ten miles without stopping," he was fond of saying, "I can make it go around the globe."

And the role that the telegraph might have played in his private life a few years earlier was all too clear to Morse. According to his son, Edward, "He recalled the days and weeks of anxiety when he was hungry for news of his loved ones; he foresaw that in affairs of state and commerce, rapid communication might mean the avoidance of war or the saving of a fortune; that, in affairs nearer to the heart of the people, it might bring a husband to the bedside of a dying wife, or save the life of a beloved child; apprehend the fleeing criminal, or commute the sentence of an innocent man."

Cooke and Morse had done the impossible and constructed working electric telegraphs. Surely the world would fall at their feet. Building the prototypes, however, proved to be the easy part. Convincing people of their significance was far more of a challenge.

3.

ELECTRIC
SKEPTICS

Although the practical working of it had been
demonstrated on a small scale, the invention seemed
altogether too chimerical to be likely ever to prove of any
worth. Again and again he was pronounced a visionary,
and his scheme stigmatized as ridiculous.

—from the *NEW YORK TIMES* obituary

of Samuel Morse, 1872

THE TROUBLE WITH the electric telegraph
was that, compared to the optical tele-
graphs that had come before it, it seemed more like a con-
juring trick than a means of communication. Anyone could
see how an optical telegraph worked: Its arms or shutters
could be set in various positions, each of which corresponded to a different letter, word, or phrase. Electric
telegraphs, on the other hand, hardly did anything at all—

they either rattled away, producing meaningless dots and dashes on a strip of paper, or had needles that wiggled convulsively. What was the good of that? Both Morse in the United States and Cooke in Britain soon realized that there was only one way they were ever going to convince the skeptics: build large-scale, working systems and demonstrate their superiority to old-fashioned optical telegraphs. So each man set out to look for the money to fund a demonstration telegraph line.

Morse thought his chance had come when a proposal was put before Congress for the construction of a line of Chappe-style telegraph towers from New York to New Orleans. The secretary of the Treasury was asked to prepare a report "upon the propriety of establishing a system of telegraphs for the United States," and a circular was issued to government officials and other interested parties asking for comments. Morse eagerly replied, explaining the advantages of an electrical design and pointing out that he had successfully sent messages through ten miles of cable. He requested funding for a demonstration network to show his telegraph working over a reasonable distance.

In 1838, Morse traveled to Washington with his apparatus and demonstrated it to government officials, but they were far from won over. It's not hard to see why: He set up his equipment on a desk, with sending and receiving stations only a few feet apart, and a huge coil of wire in between, and his talk of dots and dashes and codes didn't seem to have much to do with sending messages from one place to another. And by this time Congress seemed to have lost interest in the telegraph scheme.

Morse went to Europe in 1838–39 to popularize his invention and obtain patents for it there. In Britain, he crossed swords with Cooke and Wheatstone, but when it became clear he had no chance of being granted a British patent in the face of their objections, Morse moved on to continental Europe, where he spent several fruitless months trying to attract support.

COOKE AND WHEATSTONE were only a little more successful. Cooke's father was a friend of Francis Ronalds, whose telegraphic experiments a few years earlier had been rejected by the Admiralty. So Cooke knew that he was unlikely to get anywhere if he took his new invention to the British government. Instead, he identified a niche market for his product: the railway companies. After successfully demonstrating their apparatus to officials of the London & Birmingham Railway in 1837, Cooke and Wheatstone built an experimental telegraph link between Euston and Camden Town stations, a distance of a mile and a quarter, which worked well and boded well for the future. Cooke even drew up plans for a telegraph system linking London to Birmingham, Manchester, Liverpool, and Holyhead, which would be made available for public use. But the railway company suddenly went cold on the idea and said it "did not intend to proceed further at present."

Cooke turned to the Great Western Railway, which eventually agreed to a thirteen-mile telegraph link between Paddington and West Drayton, based on the five-needle system. Soon afterward, another telegraph system was installed on the Blackwall Railway, a small line in London's docklands. The story goes that when some of the connecting wires broke, preventing three of the five needles from working, the operators quickly improvised a new code, based on multiple wiggles, which only required two needles. At any rate, Cooke and Wheatstone soon realized there was no need for all five, which meant that subsequent installations would require fewer wires and would be much cheaper.

But the personal rivalry between the two men over which of them was principally responsible for the invention of the telegraph had, by this time, resurfaced. They eventually decided upon a gentlemanly way to resolve the matter: They appointed a panel of two mutual friends to act as arbitrators and agreed to be bound by their decision. In April 1841, the arbitrators came up with an artful compromise acceptable to both sides: "Whilst Mr. Cooke is entitled to stand alone, as the gentleman to whom this country is indebted for having practically introduced and carried out the electric telegraph as a useful undertaking, Professor Wheatstone is acknowledged as the scientific man whose profound and successful researches have already prepared the public to receive it as a project capable of practical application." In other words, the panel declined to rule in favor of either man. Almost immediately, the bickering started again.

Meanwhile, Cooke was planning to extend the Great Western Railway telegraph, but the company appeared to be losing interest. So Cooke offered to take on the running of the lines himself. He brokered a deal whereby he extended the line eighteen miles to Slough, this time with a two-needle telegraph, and could make the telegraph available to the public, on the condition that railway messages were carried for free. By this stage he had spent hundreds of pounds of his father's money for very little return. "At the beginning of 1843 we were at our lowest point of depression," he later wrote in his memoirs.

B Y T H E T I M E M O R S E got back to the United States, having failed to make any progress with his telegraph in Europe, Congress had still not got anywhere with its telegraph proposals, and his associates Gale and Vail were starting to worry that they had backed a losing horse. But Morse stubbornly refused to give up. He wrote to Vail explaining that the failure of the telegraph to take off "is not the fault of the invention, nor is it my neglect." In December 1842, he journeyed alone to Congress in a final bid for funding. He strung wires between two committee rooms in the Capitol and sent messages back and forth—and, for some reason, this time a numer of people believed him, and a bill was finally proposed allocating $30,000 toward building an experimental line.

But not everyone was convinced. As Morse watched from the gallery, Representative Cave Johnson of Tennessee ridiculed the proposal, saying that Congress might as well start funding research into mesmerism. Another skeptic joked that he had no objection to mesmeric experiments, as long as they were performed on Mr. Johnson. Amid the laughter, an amendment was put forward allocating a half share of the $30,000 to a Mr. Fisk, a well-known proponent of mesmerism. This amendment was, fortunately, rejected, and two days later the bill was passed by a vote of eighty-nine to eighty-three—a narrow margin which reflected the widespread unease that the electric telegraph might still turn out to be nothing more than an elaborate conjuring trick. But seventy congressmen chose not to vote at all, "to avoid the responsibility of spending the public money for a machine they could not understand."

Even though Morse now had the money, he still had to overcome this skepticism. He set up his apparatus again and demonstrated the transmission of messages over a five-mile-long coil of wire to any congressmen who would come and witness it. But this failed to convince them. On one occasion he transmitted the message "MR. BROWN OF

INDIANA IS HERE" down the line, walked over to the receiving apparatus, and proudly held up the strip of paper with the message spelled out in dots and dashes. "It won't do. That doesn't prove anything," whispered one onlooker. "That's what I call pretty thin," said another. Senator Oliver Smith of Indiana, who attended one of Morse's demonstrations, recalled that he "watched his countenance closely, to see if he was not deranged . . . and I was assured by other senators after we left the room that they had no confidence in it."

Morse decided to press ahead all the same with a line from Washington to Baltimore, a distance of about forty miles. The two towns were already linked by railway, and he obtained permission to run the telegraph cable alongside the railway. The Baltimore & Ohio Railroad Company was more than a little suspicious; it granted permission on the condition that the line could be built "without embarrassment to the operations of the company" and, just to cover itself both ways, demanded free use of the telegraph if indeed it did turn out to work.

In the spring of 1844, an observer, John W. Kirk, was appointed by Congress to keep an eye on Morse, who was described as "impracticable or crazy" and whose invention was regarded as mere "foolishness." But although he started out as skeptical as everyone else, Kirk soon came up with a scheme that would verify whether or not all those dots and dashes actually corresponded to useful information. The Whig National Convention was due to take place in Baltimore on May 1, and although the line had not been completed by then, it did reach from Washington to within fifteen miles of Baltimore. By successfully transmitting the names of the convention's nominees, Morse would be able to prove the usefulness of his invention.

Once announced, the names of the nominees were duly rushed by train to Vail, who was installed at a temporary platform fifteen miles outside Baltimore. Vail then transmitted the list to Morse in Washington, where a large crowd had gathered at the railroad depot. The names were announced to the crowd, and were confirmed when the first train arrived from Baltimore sixty-four minutes later—forcing even the staunchest skeptics to eat their words.

The line was soon completed to Baltimore, and on May 24, 1844, Morse officially inaugurated it by transmitting a message from the Supreme Court chamber in Washington to Vail in Baltimore: "WHAT HATH GOD WROUGHT." The wonders of the electric telegraph were written up in the newspapers, and Morse's success seemed assured.

Yet after a while he realized that everybody still thought of the telegraph as a novelty, as nothing more than an amusing subject for a newspaper article, rather than the revolutionary new form of communication that he envisaged.

In BRITAIN, Cooke had by this time licensed the use of the Paddington-Slough line to a promoter, Thomas Home, who opened it up for public demonstration. The poster printed to advertise this new attraction says a lot about the way it was perceived: "Despatches sent instantaneously to and fro with the most confiding secrecy," it proclaimed. "Post Horses and Conveyances of every description may be ordered by the electric telegraph to be in readiness on the arrival of a train, at either Paddington or Slough Station." The *Morning Post* declared that the exhibition was "well worthy of a visit from all who love to see the wonders of science." Sending messages to and fro was merely thought of as a scientific curiosity; the telegraph was evidently not regarded as a useful form of communication. As the small print of the poster pointed out, "Messengers in constant attendance, so that communications received by telegraph would be forwarded, if required, to any part of London, Windsor, Eton, & c." But this was seen as incidental to the main attraction.

The fame of the telegraph took a giant leap when it was used to announce the birth of Queen Victoria's second son, Alfred Ernest, at Windsor on August 6, 1844. The *Times* was on the streets of London with the news within forty minutes of the announcement, declaring itself "indebted to the extraordinary power of the Electro-Magnetic Telegraph" for providing the information so quickly. Three trainloads of assorted lords and gentry then left London for a celebratory banquet at Windsor, and the telegraph proved its use once again. The Duke of Wellington forgot his dress suit, so he telegraphed London and asked for it to be sent on the following train. Fortunately, it arrived in time for the royal banquet.

Another milestone for the telegraph was when it was used to apprehend Fiddler Dick, a notorious pickpocket, and his gang. Their modus operandi involved robbing the crowds at a busy railway station and then escaping from the scene by train. Before the telegraph, there was no way to send information faster than a speeding train, so their getaway was assured. However, the presence of the telegraph alongside the Paddington-Slough line meant it was now possible to alert the police at the other end before the train's arrival.

More famous still was the arrest of John Tawell on January 3, 1845, thanks to the telegraph. Tawell had murdered his mistress in Slough, and when his crime was discovered he made a run for it and headed for London. He was dressed in a brown, unusual-looking greatcoat. His description, "DRESSED LIKE A KWAKER" (since Cooke and Wheatstone's telegraphic alphabet had no "Q"), was sent to London, where the police were able to meet the train and arrest Tawell before he had time to melt into the crowds. "It may be observed," reported the *Times*, "that had it not been for the efficient aid of the electric telegraph, both at Slough and Paddington, the greatest difficulty as well as delay would have occurred in the apprehension of the party now in custody." Tawell was subsequently convicted and hanged, and the telegraph wires gained further notoriety as "the cords that hung John Tawell."

All of this enabled Thomas Home to print up a new edition of his poster. Now the telegraph was described as "an Exhibition admitted by its numerous visitors to be the most interesting and attractive of any in this great Metropolis. In the list of visitors are the illustrious names of several of Crowned Heads of Europe, and nearly the whole of the Nobility of England." Home had evidently devised new ways to demonstrate the speed of the telegraph and made the most of its crime-fighting capabilities: "Questions proposed by visitors will be asked by means of this apparatus and answers thereto will instantaneously be returned by a person 20 miles off, who will also, at their request, ring a bell or fire a cannon, in an incredibly short space of time, after the signal for his so doing so has been given. By its powerful agency murderers have been apprehended, thieves detected, and lastly, which is of no little importance, the timely assistance of medical aid has been procured in cases which otherwise would have proved fatal. The great national importance of this wonderful invention is so well known that any further allusion here to its merits would be quite superfluous." Once again, the telegraph's potential use as a means of sending useful messages between Slough and London was buried in the small print at the bottom of the poster.

In THE UNITED STATES, Morse and his associates faced similar apathy. Even though the use of the experimental Washington-Baltimore telegraph was free, members of the public were quite content just to come and see it, and watch chess games played between the leading players of each town over the wires. But the telegraph wasn't regarded as being useful in day-to-day life. "They would not say a word or stir and didn't care whether they understood or not, only they wanted to say they had seen it," Vail complained to Morse.

Before long, religious leaders in Baltimore expressed their doubts about the new technology, which was too much like black magic for their liking, prompting Henry J. Rogers, the Baltimore operator, to warn Vail that "if we continue we will be injured more than helped." Aware of the importance of keeping public opinion on their side, they decided to call a halt to the frivolous chess games and restrict use of the line to congressional business.

In June 1844, Morse went back to Congress to press for the extension of the line from Baltimore to New York. He presented to the House several examples of the benefits of the telegraph. A family in Washington, for instance, had heard a rumor that one of their relatives in Baltimore had died, and asked Morse to find out if it was true. Within ten minutes they had their answer: The rumor was false. Another example concerned a Baltimore merchant who telegraphed the Bank of Washington to verify the creditworthiness of a man who had written him a check. But

Congress still adjourned for the summer without making a decision. In December, Morse appealed to the House again, pointing out that the telegraph would be much more useful with more stations, and advocating the wiring up of the country's major cities.

Again, he gave examples of the benefits of the Washington-Baltimore line; in a case similar to that of John Tawell, police in Baltimore had been able to arrest a criminal as he stepped from an arriving train, after his description was telegraphed to them by the police in Washington. By this stage, the proceedings of Congress were being transmitted for inclusion in the Baltimore papers, and one or two farsighted businessmen were starting to use the line. But again, nothing happened.

Morse, disheartened by the government's lack of interest, turned to private enterprise. He teamed up with Amos Kendall, a former politician and journalist, whom he appointed as his agent. Kendall proposed the construction of lines along major commercial routes radiating out from New York using private money, with Morse and the other patent holders to be awarded 50 percent of the stock of each telegraph company formed in return for the patent rights. In May 1845, the Magnetic Telegraph Company was formed, and by the autumn, lines were under construction toward Philadelphia, Boston, and Buffalo, and westward toward the Mississippi.

Meanwhile, the postmaster general, Cave Johnson, the former congressman who had ridiculed the whole idea of the telegraph two years earlier, decided that the time had come for the government to try to get some of its money back from the Washington-Baltimore telegraph. He imposed a tariff of one cent for four characters, and on April 1, 1845, the line was officially opened for public business. It was a financial failure: The line took one cent during its first four days of operation. A man with only a twenty-dollar bill and one cent entered the Washington office and asked for a demonstration, so Vail offered him a half cent's worth of telegraphy: He sent to Rogers in Baltimore the digit 4, which was short for "WHAT TIME IS IT?" and the answer, the digit 1, meaning one "O'CLOCK," came back. This was hardly an impressive demonstration, since it was the same time in both cities, and the customer left without even asking for his half cent change.

During the fifth day the telegraph took twelve and a half cents, and revenues rose slowly to reach $1.04 by the ninth day—hardly big business. After three months the line had taken in $193.56 but had cost $1,859.05 to run. Congress decided to wash its hands of the whole affair and handed the line over to Vail and Rogers, who agreed to maintain the line at their own expense in return for the proceeds. In light of this development, Kendall's optimism that the telegraph would take off seemed unfounded.

But he knew what he was doing. By the following January, the Magnetic Telegraph Company's first link was completed between New York and Philadelphia, and Kendall placed advertisements in the newspapers at both ends announcing the opening of the line on January 27. The fee for a message was set at twenty-five cents for ten words.

The receipts for the first four days were an impressive $100. "When you consider," wrote the company treasurer, "that business is extremely dull, we have not yet the confidence of the public and that on two of the four days we have been delayed and lost business in Philadelphia through mismanagement, you will see we are all well satisfied with results so far. In one month we shall be doing a $50 business a day."

In BRITAIN, the tide had also turned for Cooke, who had scored a notable victory: He had actually persuaded the Admiralty to sign a valuable contract for an eighty-eight-mile electric telegraph link between London and Portsmouth. Clearly, if he could convince the Admiralty, he could convince anyone. The success of the line led to more business, and proposals were advanced for a telegraph linking London with the key industrial centers of Manchester, Birmingham, and Liverpool—which would have obvious commercial applications. Cooke had also signed up more railway companies, and hundreds of miles of line were soon under construction.

The arrest of Tawell had brought the electric telegraph to the attention of John Lewis Ricardo, a member of Parliament and a prominent financier. He purchased a share of the patent rights to the telegraph from Cooke and

Wheatstone, clearing Cooke's debts and valuing their business at a hefty £144,000. In September 1845, Cooke and Ricardo set up the Electrical Telegraph Company, which bought out Cooke and Wheatstone's patent rights altogether.

On both sides of the Atlantic, the electric telegraph was finally taking off.

4.

THE THRILL ELECTRIC

"We are one!" said the nations, and hand met hand,
in a thrill electric from land to land.

> —from "The Victory," a poem written in tribute to
> Samuel Morse, 1872

No invention of modern times has extended its influence so rapidly as that of the electric telegraph," declared *Scientific American* in 1852. "The spread of the telegraph is about as wonderful a thing as the noble invention itself."

The growth of the telegraph network was, in fact, nothing short of explosive; it grew so fast that it was almost impossible to keep track of its size. "No schedule of telegraphic lines can now be relied upon for a month in succession," complained one writer in 1848, "as hundreds of miles may be added in that space of time. It is anticipated that the whole of the populous parts of the United States will, within two or three years, be covered with net-work like a spider's web."

Enthusiasm had swiftly displaced skepticism. The technology that in 1845 "had been a scarecrow and chimera, began to be treated as a confidential servant," noted a report compiled by the Atlantic and Ohio Telegraph Company in 1849. "Lines of telegraph are no longer experiments," declared the *Weekly Missouri Statesman* in 1850.

Expansion was fastest in the United States, where the only working line at the beginning of 1846 was Morse's experimental line, which ran 40 miles between Washington and Baltimore. Two years later there were approximately 2,000 miles of wire, and by 1850 there were over 12,000 miles operated by twenty different companies. The telegraph industry even merited twelve pages to itself in the 1852 U.S. Census.

"The telegraph system [in the United States] is carried to a greater extent than in any other part of the world," wrote the superintendent of the Census, "and numerous lines are now in full operation for a net-work over the length and breadth of the land." Eleven separate lines radiated out from New York, where it was not uncommon for some bankers to send and receive six or ten messages each day. Some companies were spending as much as $1,000 a year on telegraphy. By this stage there were over 23,000 miles of line in the United States, with another

10,000 under construction; in the six years between 1846 and 1852 the network had grown 600-fold.

"Telegraphing, in this country, has reached that point, by its great stretch of wires and great facilities for transmission of communications, as to almost rival the mail in the quantity of matter sent over it," wrote Laurence Turnbull in the preface to his 1852 book, *The Electro-Magnetic Telegraph*. Hundreds of messages per day were being sent along the main lines, and this, wrote Turnbull, showed "how important an agent the telegraph has become in the transmission of business communications. It is every day coming more into use, and every day adding to its power to be useful."

Arguably the single most graphic example of the telegraph's superiority over conventional means of delivering messages was to come a few years later, in October 1861, with the completion of the transcontinental telegraph line across the United States to California. Before the line was completed, the only link between East and West was provided by the Pony Express, a mail delivery system involving horse and rider relays. Colorful characters like William "Buffalo Bill" Cody and "Pony Bob" Haslam took about 10 days to carry messages over the 1,800 miles between St. Joseph, Missouri and Sacramento. But as soon as the telegraph line along the route was in place, messages could be sent instantly, and the Pony Express was closed down.

In Britain, where the telegraph was doing well but had

Construction of the transcontinental telegraph along the route of the Pony Express, 1861. When the telegraph line was complete, the horse-and-rider relay service was rendered obsolete.

not been quite so rapidly embraced, there was some be-musement at the enthusiasm with which it had been adopted on the other side of the Atlantic. "The American telegraph, invented by Professor Morse, appears to be far more cosmopolitan in the purposes to which it is applied than our telegraph," remarked one British writer, not without disapproval. "It is employed in transmitting messages to and from bankers, merchants, members of Congress, officers of government, brokers, and police officers; parties who by agreement have to meet each other at the two stations, or have been sent for by one of the parties; items of news, election returns, announcements of deaths, inquiries respecting the health of families and individuals, daily proceedings of the Senate and the House of Representatives, orders for goods, inquiries respecting the sailing of vessels, proceedings of cases in various courts, summoning of witnesses, messages for express trains, invitations, the receipt of money at one station and its payment at another; for persons requesting the transmission of funds from debtors, consultation of physicians, and messages of every character usually sent by the mail. The confidence in the efficiency of telegraphic communication has now become so complete, that the most important commercial transactions daily transpire by its means between correspondents several hundred miles apart."

Just as the old optical telegraphs were understood to be the preserve of the Royal Navy, the new electric telegraph was associated in British minds with the railways. By 1848, about half of the country's railway tracks had telegraph wires running alongside them. By 1850, there were 2,215 miles of wire in Britain, but it was the following year that things really took off. The domination enjoyed by Ricardo and Cooke's Electric Telegraph Company came to an end as rival companies arrived on the scene, and thirteen telegraph instruments based on a variety of designs were displayed at the Great Exhibition of 1851 in London, fueling further interest in the new technology. These developments gave the nascent industry the jolt it needed to emerge from the shadow of the railways.

The telegraph was doing well in other countries, too. By 1852, there was a network of 1,493 miles of wire in

Prussia, radiating out from Berlin. Turnbull, who compiled a survey of telegraph systems around the world, noted that instead of stringing telegraph wires from poles, "the Prussian method of burying wires beneath the surface protects them from destruction by malice, and makes them less liable to injury by lightning." Austria had 1,053 miles of wire, and Canada 983 miles; there were also electric telegraphs in operation in Tuscany, Saxony and Bavaria, Spain, Russia, and Holland, and networks were being established in Australia, Cuba, and the Valparaiso region of Chile. Competition thrived between the inventors of rival telegraph instruments and signaling codes as networks sprung up in different countries and the technology matured.

Turnbull was pleased to note that the wonders of the telegraph had managed to rouse the "lethargic" inhabitants of India into building a network. He was even ruder about the French, whom he described as "inferior in telegraphic enterprise to most of the other European companies." This view was unfounded, for the French had not only invented the telegraph but named it too. But their lead in the field of optical telegraphy had actually worked against them, and the French were reluctant to abandon the old technology in favor of the new. François Moigno, a French writer, compiled a treatise on the state of the French electric telegraph network, whose size he put at a total of 750 miles in 1852—and which he condemned for leading to the demise of the old optical telegraphs.

ENDING AND RECEIVING messages—
which by the early 1850s had been dubbed
"telegrams"—soon became part of everyday life for many
people around the world. But because this service was expensive, only the rich could afford to use the network to
send trivial messages; most people used the telegraph
strictly to convey really urgent news.

Sending a message was a matter of going into the office of one of the telegraph companies and filling in a form
giving the postal address of the recipient and a message—
expressed as briefly as possible, since messages were
charged by the word, as well as by the distance from
sender to receiver. Once the message was ready to go, it
would be handed to the clerk, who would transmit it up
the line.

Telegraph lines radiated out from central telegraph offices in major towns, with each line passing through several local offices, and long-distance wires linking central
offices in different towns. Each telegraph office could only
communicate with offices on the same spoke of the network, and the central telegraph office at the end of the
line. This meant that messages from one office to another
on the same spoke could be transmitted directly, but that
all other messages had to be telegraphed to the central
office and were then retransmitted down another spoke of
the network toward their final destination.

Once received at the nearest telegraph office, the message was transcribed on a paper slip and taken on foot by a messenger boy directly to the recipient. A reply, if one was given, would then be taken back to the office; some telegraph companies offered special rates for a message plus a prepaid reply.

Young men were eager to enter the business as messengers, since it was often a stepping-stone to better things. One of the duties of messenger boys was to sweep out the operating room in the mornings, and this provided an opportunity to tinker on the apparatus and learn the telegrapher's craft. Thomas Edison and steel magnate and philanthropist Andrew Carnegie both started out as telegraph messenger boys. "A messenger boy in those days had many pleasures," wrote Carnegie in his autobiography, which includes rather rose-tinted reminiscences of the life of a messenger boy in the 1850s. "There were wholesale fruit stores, where a pocketful of apples was sometimes to be had for the prompt delivery of a message; bakers and confectioners' shops where sweet cakes were sometimes given to him. He met very kind men to whom he looked up with respect; they spoke a pleasant word and complimented him on his promptness, perhaps asking him to deliver a message on the way back to the office. I do not know a situation in which a boy is more apt to attract attention, which is all a really clever boy requires in order to rise."

Though its business was the sending and receiving of messages, much like e-mail today, the actual operation of the telegraph had more in common with an on-line chat room. Operators did more than just send messages back and forth; they had to call up certain stations, ask for messages to be repeated, and verify the reception of messages. In countries where Morse's apparatus was used, skilled operators quickly learned to read incoming messages by listening to the clicking of the apparatus, rather than reading the dots and dashes marked on the paper tape, and this practice soon became the standard means of receiving. It also encouraged more social interaction over the wires, and a new telegraphic jargon quickly emerged.

Rather than spell out every word ("PHILADELPHIA CALLING NEW YORK") letter by letter in laborious detail, conventions arose by which telegraphers talked to each other over the wires using short abbreviations. There was no single standard: different dialects or customs arose on different telegraph lines. However, one listing of common abbreviations compiled in 1859 includes "I I" (dot dot, dot dot) for "I AM READY"; "G A" (dash dash dot, dot dash) for "GO AHEAD"; "S F D" for "STOP FOR DINNER"; "G M" for "GOOD MORNING." This system enabled telegraphers to greet one another and handle most common situations as easily as if they were in the same room. Numbers were also used as abbreviations: 1 meant "WAIT A MOMENT"; 2, "GET ANSWER IMMEDIATELY"; 33, "ANSWER PAID HERE." All telegraph offices on a branch line shared one wire, so at any time there could be several telegraphers listening in to wait for the line to become available. They could also chat, play chess, or tell jokes during quiet periods.

ALTHOUGH THE TELEGRAPH, unlike later forms of electrical communication, did not require the consumer who was sending or receiving a message to own any special equipment—or understand how to use it—it was still a source of confusion to those unfamiliar with it. And just like the apocryphal story of the woman who tried to send her husband tomato soup by pouring it into the telephone handset, there are numerous stories of telegraph-inspired confusion and misunderstanding.

One magazine article, "Strange Notions of the Telegraph," gives several examples of incomprehension: "One wiseacre imagined that the wires were hollow, and that papers on which the communications were written were blown through them, like peas through a pea shooter. Another decided that the wires were speaking tubes." And one man in Nebraska thought the telegraph wires were a kind of tightrope; he watched the line carefully "to see the man run along the wires with the letter bags."

In one case a man came into a telegraph office in Maine, filled in a telegraph form, and asked for his message to be sent immediately. The telegraph operator tapped it out in Morse to send it up the line and then spiked the form on the "sent" hook. Seeing the paper on the hook, the man assumed that it had yet to be transmitted. After waiting a few minutes, he asked the telegrapher, "Aren't you going to send that dispatch?" The operator explained that he already had. "No, you haven't," said the man, "there it is now on the hook."

Another story concerned a woman in Karlsruhe, Prussia, who went to a telegraph office in 1870 with a dish full of sauerkraut, which she asked to have telegraphed to her son, who was a soldier fighting in the war between Prussia and France. The operator had great difficulty convincing her that the telegraph was not capable of transmitting objects. But the woman insisted that she had heard of soldiers being ordered to the front by telegraph. "How could so many soldiers have been sent to France by telegraph?" she asked.

As one magazine article of the time pointed out, much confusion resulted from the new electric jargon, which imposed new meanings on existing words. "Thus, when it is said that a current of electricity flows along a wire, that the wire or the current carries a message, the speaker takes language universally understood, relating to a fluid moving from one place to another, and a parcel or a letter transported from place to place." One young girl asked her mother how the messages "get past the poles without being torn." The mother is said to have replied, "They are sent in a fluid state, my dear."

And there was a widespread belief that it was possible to hear the messages as they passed along the wires. According to a book, *Anecdotes of the Telegraph*, published in 1848, "a very general but erroneous idea, even among the better order of folks, is that the humming aeolian harp-like effect of the wind on the suspended wire is caused by the messages passing." A typical story concerned a telegraph operator who worked in a station in the Catskill Mountains, where the wind often whistled through the wires. One day a local man asked how business was doing. "Lively," said the operator. "Well, I didn't think so," said the man, "I ain't heard a dispatch go up in three or four days."

The retranscription of the message at the receiving station also confused some people. One woman preparing to send a telegram is said to have remarked as she filled out the telegraph form, "I must write this out afresh, as I don't want Mrs. M. to receive this untidy telegram." Another woman, on receiving a telegram from her son asking for money, said she was not so easily taken in; she knew her son's handwriting very well, she said, and the message, transcribed at the receiving office, obviously hadn't come from him.

a S TELEGRAPH NETWORKS sprung up in different countries, the benefits of joining them soon became apparent. The first interconnection treaty was signed on October 3, 1849, between Prussia and Austria, so that messages could be sent from Vienna to Berlin. It was an inefficient system; rather than running a wire across the border, a special joint telegraph office was constructed, staffed by representatives of each country's telegraph company, who were connected to their respective national networks. When a message needed to be passed from one country to another, it was transcribed by the clerks at one end of the office, who then physically handed it over to their opposite numbers at the other end of the office for retransmission.

Similar agreements were soon in place between Prussia and Saxony, and Austria and Bavaria. In 1850, the four states established the Austro-German Telegraph Union to regulate tarriffs and set common rules for interconnection. The following year, the Morse telegraph system was adopted as a standard to allow direct connections to be established between the four networks. Soon interconnection agreements had also been signed between France, Belgium, Switzerland, Spain, and Sardinia. But if Britain was to be connected to the growing European network, a significant barrier would have to be overcome: the English Channel.

Actually, experiments with sending messages along underwater telegraph cables had been going on almost since the earliest days of electric telegraphy. Wheatstone had tried it out in Wales, sending messages from a boat to a lighthouse, and in 1840 he proposed the establishment of a cross-Channel telegraph. But at that time the telegraph had yet to prove itself over short distances on land, let alone across water.

Morse, too, had a go at underwater telegraphy. In 1843, after coating a wire in rubber and encasing it in a lead pipe, he sent messages along a submerged cable between Castle Garden and Governors Island in New York

Harbor. He also succeeded in using water itself as the conductor, with metal plates dipped in the water on each bank of a river and connected to the telegraph wires. (Wheatstone did some similar experiments across the river Thames in the presence of Prince Albert the same year.) At any rate, Morse was sufficiently pleased with the results across a few feet of water that, in typical indefatigable Morse fashion, he predicted that it wouldn't be long before there would be telegraph wires across the Atlantic.

For advocates of cross-Channel telegraphy, however, there were practical problems to be overcome. Laying a rubber-coated wire inside a lead pipe was possible in New York Harbor; laying a pipe along the seabed across the English Channel was another matter entirely. And if the cable was to last any length of time, an alternative to coating it in rubber would have to be found, since rubber quickly deteriorated in water.

The solution was to use gutta-percha, a kind of rubbery gum obtained from the gutta-percha tree, which grows in the jungles of Southeast Asia. One useful property of gutta-percha is that it is hard at room temperature but softens when immersed in hot water and can be molded into any shape. The Victorians used it much as we use plastic today. Dolls, chess pieces, and ear trumpets were all made of gutta-percha. And although it was expensive, it turned out to be ideal for insulating cables.

Once the question of what to use for insulation had been resolved, John Brett, a retired antique dealer, and his younger brother Jacob, an engineer, decided to embark upon building a telegraph link between England and France. They got the appropriate permission from the British and French governments and ordered a wire coated with a quarter of an inch of gutta-percha from the Gutta Percha Company in London. Their plan was breathtakingly low-tech: They intended to spool the wire (which was about the thickness of the power cable of a modern domestic appliance) out of the stern of a boat as it steamed across the Channel. They would then connect telegraph instruments at each end, and their company, the grandly named General Oceanic and Subterranean Electric Printing Telegraph Company, would be in business. On August 28, 1850, with their cable wound onto a vast drum and mounted on the back of a small steam tug, the *Goliath*, they set out for France.

Things did not go according to plan. For starters, the wire was so thin that it wouldn't sink; it simply floated pathetically in the water behind the boat. The Bretts' response was to clamp weights around the wire at regular intervals to get it to sink. By the evening, they had arrived at Cap Gris-Nez near Calais in France, where they wired up their newfangled telegraph instrument—the very latest automatic printing model—and waited for the first test message to be sent from England. It came out as gibberish.

The cable was working, but the messages were being garbled because the surrounding water changed the cable's electrical properties in a way that was poorly understood at the time. Effectively, it meant that the staccato pulses of electricity were smoothed out, and the Bretts' high-speed automatic machines transmitted so fast that succeeding pulses overlapped and became indistinct. But, using an old-fashioned single-needle telegraph, they were eventually able to send a few messages manually, in much the same way that a preacher in a resonant cathedral must speak slowly and distinctly in order to be understood. However, the next day the cable met a watery end; a French fisherman snagged it in his net, and when he brought it to the surface he hacked off a piece to see what it was. Deciding that it was a hitherto unknown form of seaweed with a gold center, he took it to show his friends in Boulogne.

It took the Bretts over a year to raise the money for another cable, and they would probably have had to give up altogether but for the intervention of Thomas Crampton, a railway engineer. He put up half the £15,000 needed, and also designed the new cable. He wanted to protect his investment, so the new cable consisted of four gutta-percha-covered wires twisted together and wrapped in tar-covered hemp, and then encased in a cladding of tar-covered iron cords. It was far tougher than the first cable, and it weighed thirty times as much. This meant it was harder to lay—not because it wouldn't sink, like the first cable, but because it was so heavy it ran off the drum on the back of the boat faster than the Bretts wanted it to. It was so hard to control, in fact, that all the cable had been paid out before the boat carrying it reached France. Fortunately, the Bretts had brought along a spare piece of cable, which they spliced on, and in November 1851, after a few weeks of testing, the cable was opened to the public. The first direct message from London to Paris was sent in 1852.

The success of the Channel cable led to a boom in submarine telegraphy—to the delight of the directors of the Gutta Percha Company. With a virtual monopoly on the supply of gutta-percha, they suddenly found they were sitting on a gold mine. The problem of laying a telegraph link across a stretch of water seemed to have been cracked: It was simply a matter of making sure that the cable was properly insulated, strong enough not to break, and heavy enough to sink, and that messages weren't sent too quickly along it. Before long Dover had been linked to Ostend, and after two failed attempts England was linked to Ireland in 1853. Further underwater links across the North Sea directly connected Britain with the coasts of Germany, Russia, and Holland. John Brett soon turned his attention to linking Europe with Africa and succeeded in connecting Corsica and Sardinia to Genoa on the European mainland in 1854. But the following year, he failed in his attempt to reach the North African coast, which involved laying a cable across the deepest and most mountainous part of the Mediterranean seabed. Brett lost a lot of money, and his failure proved that there were limits to submarine telegraphy after all. The prospect of linking Europe and North America seemed as far away as ever.

5.

WIRING
THE WORLD

The Atlantic Telegraph—that instantaneous highway of
thought between the Old and New Worlds.
—*SCIENTIFIC AMERICAN*, 1858

T HE IDEA OF a transatlantic telegraph had
been mooted by Morse and others since the
1840s, but, much as we regard time machines or interstellar travel today, in the 1850s it was generally regarded as
something that was very unlikely ever to come to pass—
though it would certainly have its uses if it did.

The difficulties facing a transatlantic telegraph were
obvious. "Fancy a shark or a swordfish transfixing his fins
upon the insulated wires, in the middle, perhaps, of the
Atlantic, interrupting the magic communication for
months," wrote one skeptic. "What is to be done against
the tides, when they deposit their floating debris of wrecks

and human bodies? Even supposing you could place your wires at the lowest depth ever reached by plumb line, would your wires, even then, be secure?"

Nobody who knew anything about telegraphy would be foolish enough to risk building a transatlantic telegraph; besides, it would cost a fortune. So it's hardly surprising that Cyrus W. Field, the man who eventually tried to do it, was both ignorant of telegraphy and extremely wealthy. He was a self-made man from New England who amassed his fortune in the paper trade and retired at the age of thirty-three. After spending a few months traveling, he happened to meet an English engineer, Frederic N. Gisborne, who introduced him to the business of telegraphy.

Gisborne was looking for a backer after his failed attempt in 1853 to build a telegraph cable across Newfoundland, with a link to the mainland across the Gulf of St. Lawrence. His plan had been sound enough: Since building a cable all the way across the Atlantic seemed both technically and financially out of the question, building a link from New York to St. John's, on the eastern tip of Newfoundland, was the next best thing. Steamers could stop at St. John's on their way westward, and messages could then be forwarded by telegraph to New York, reducing the time taken for messages to arrive from Europe by several days.

The trouble was, Gisborne's plan involved stringing a cable across some of the coldest, most inhospitable terrain on earth. And even with the use of four local guides—of which two ran away and one died—he was forced to abandon his first attempt after only a few miles of cable had been laid. So when he visited Field in January 1854, Gisborne hoped to convince him that telegraphy was a worthwhile business to invest in. Evidently he did a very good job because, according to Field's brother Henry, as soon as the meeting with Gisborne was over, Field "went to the globe in his library and began to turn it over." He soon set his heart on a much grander scheme—building a cable right across the Atlantic. Newfoundland would be merely one of the stops along the way.

Confident that he would be able to handle the business side of things, Field wanted to make sure that there were no technical barriers standing in his way. He wrote to Morse to inquire about the feasibility of a cable from Newfoundland to Europe. At the same time, he wrote to Matthew Fontaine Maury, the leading hydrographer in the United States. Maury had compiled readings from the logs of hundreds of ships into the most accurate charts of the Atlantic in existence, so he was the logical person to suggest the route for the cable. Oddly enough, his charts had revealed the presence of a large raised plateau on the seabed between Newfoundland and Ireland, which Maury had already realized would be ideal for "holding the wires of a submarine telegraph and keeping them out of harm's way." Morse, who wanted to see his prediction of an Atlantic telegraph fulfilled, also gave his backing to the scheme, and before long Field had resurrected Gisborne's old company and set about the construction of the line across Newfoundland.

After two and a half years of hard work, the New York–St. John's link was complete. By this time, Field had established the New York, Newfoundland and London Telegraph Company, and his next step was to go to London and drum up support for the cable on the other side of the Atlantic. There he met John Brett, who was eager to get involved; Morse was also in London at the time, conducting an important experiment. After connecting ten telegraph lines, each of which ran the 200 miles from London to Manchester, Morse was able to send signals sucessfully around the whole circuit. This suggested that telegraphy would indeed be possible over a 2,000-mile cable, and ensured that there were plenty of investors ready to support the new company Field and Brett were establishing in London.

The Atlantic Telegraph Company was duly set up, and Field persuaded the British and United States governments to back his project; in return for an annual subsidy and the provision of ships to help lay the cable, official messages would be carried free of charge. Construction of the 2,500-mile cable began, following the precise specifications of the company's newly appointed official electrician, Dr. Edward Orange Wildman Whitehouse. Everything seemed to be going according to plan. There was only one fly in the ointment: Whitehouse was totally incompetent.

THE FACT THAT THE JOB of designing the world's longest telegraph cable was given to an amateur like Whitehouse shows how little the scientific understanding of telegraphy had advanced in the previous twenty years. Whitehouse had started out as a surgeon and had taught himself everything he knew about telegraphy—which wasn't very much. In some fields, practical experience is every bit as valuable as theoretical understanding, but Whitehouse had neither, even though he had spent years experimenting with telegraphic equipment. Field, who knew nothing about technical matters, liked the fact that Whitehouse didn't bother with theory and instead trusted in his own experimental results. And since Field was running the show, the doctor got the job. Whitehouse then managed to get nearly every aspect of the cable's design wrong.

In particular, his experiments led him to conclude that messages should be sent over the cable using high voltages generated by huge induction coils and that the conducting wire should have a small rather than a large diameter. Whitehouse claimed that "no adequate advantage would be gained by any considerable increase in the size of the wire." Unfortunately for the cable's backers, he was mistaken on both counts. To make matters worse, Field had promised that the Atlantic telegraph would start operating by the end of 1857, and he was in such a desperate hurry that the manufacture of the cable was rushed. As a result, parts of it didn't even come up to the inadequate standards Whitehouse had laid down.

Even so, the cable was taken to sea in July 1857. It was half an inch thick and weighed a ton per mile. Since no ship afloat could carry 2,500 tons of cable, half of it went aboard the steam frigate USS *Niagara*, the finest ship in the U.S. Navy; the other half was loaded onto the British vessel HMS *Agamemnon*. The ships, accompanied by two escorts, headed for Valentia Bay in the southwest of Ireland, which had been chosen as the best place for the cable to come ashore. The plan was for the *Niagara* to spool out its half of the cable as the fleet headed west; in the mid-Atlantic the *Agamemnon*'s half would then be joined on for the remaining half of the journey. However, after a few days, when about 350 miles of the cable had been laid, it snapped and fell into the sea.

It took Field several months to raise the money for an additional length of replacement cable and a second expedition. But in June of the following year, the ships set out again, this time with a new plan: They would sail to the middle of the Atlantic, join the two halves of the cable, and set out in opposite directions. This would, in theory, halve the time it would take to lay the cable. After weathering a particularly unpleasant storm, the fleet assembled at the halfway point, spliced the two halves of the cable, and set out in opposite directions. Twice the cable snapped, and twice they sailed back to the rendezvous and started again. The *Agamemnon* also had a close encounter with a whale. When the cable broke for the third time, the ships went back to Ireland to take on new provisions, before

HMS *Agamemnon* encounters a whale during the laying of the first transatlantic cable, July 1858. Both whale and cable emerged unscathed.

setting out once again. Eventually, on the fourth attempt, and having laid 2,050 miles of cable between them, the *Agamemnon* reached Newfoundland and the *Niagara* reached Valentia Bay. The cable was landed on August 5, 1858. For the first time, the telegraph networks of Europe and North America had been connected.

THE CELEBRATIONS that followed bordered on hysteria. There were hundred-gun salutes in Boston and New York; flags flew from public buildings; church bells rang. There were fireworks, parades, and special church services. Torch-bearing revelers in New York got so carried away that City Hall was accidentally set on fire and narrowly escaped destruction.

"Our whole country," declared *Scientific American*, "has been electrified by the successful laying of the Atlantic Telegraph." The New York newspapers of August contained, according to one writer, "hardly anything else than popular demonstrations in honor of the Atlantic Telegraph. It was indeed a national jubilee."

Field was inundated with congratulations—"an avalanche of praise"—and his acknowledgments, in which he thanks everyone else who helped make the cable possible, read like a very long Oscar acceptance speech. Queen Victoria exchanged appropriately epic messages over the cable with President James Buchanan, who described it as "a triumph more glorious, because far more useful to mankind, than was ever won by a conqueror on the field of battle." And a lot of extraordinarily bad poetry was written.

> 'Tis done! The angry sea consents,
> the nations stand no more apart;
> with clasped hands the continents,
> feel the throbbings of each other's hearts.
> Speed, speed the cable, let it run,
> a loving girdle round the earth,
> till all the nations neath the sun
> shall be as brothers of one hearth.

Suitably telegraphic biblical references were un-
earthed by preachers, notably "Their line is gone out
through all the earth, and their words to the end of the
world" (Psalms 19) and "Canst thou send lightnings that
they may go, and say unto Thee, here we are?" (Job 38).

Tiffany's, the New York jewelers, bought the remain-
der of the cable, cut it into four-inch pieces, and sold
them as souvenirs. Pieces of spare cable were also made
into commemorative umbrella handles, canes, and watch
fobs. "Nothing seemed too extravagant to give expression
to the popular rejoicing," Henry Field later wrote in a bi-
ography of his brother.

Books explaining the construction and working of the
cable were rushed out to capitalize on the sudden interest
in all things telegraphic. "The completion of the Atlantic
telegraph, the unapproachable triumph which has just
been achieved, has been the cause of the most exultant
burst of popular enthusiasm that any event in modern
times has ever elicited," wrote Charles Briggs and Augustus Maverick in their hastily compiled tome, *The Story
of the Telegraph*. "The laying of the telegraph cable is regarded, and most justly, as the greatest event in the present century; now the great work is complete, the whole
earth will be belted with electric current, palpitating with
human thoughts and emotions. It shows that nothing is
impossible to man."

In London, the *Times* compared the laying of the cable

with the discovery of the New World: "Since the discovery of Columbus, nothing has been done in any degree comparable to the vast enlargement which has thus been given to the sphere of human activity." Another widely expressed sentiment, also articulated by the *Times*, was that the cable had reunited the British and American peoples: "The Atlantic is dried up, and we become in reality as well as in wish one country. The Atlantic Telegraph has half undone the Declaration of 1776, and has gone far to make us once again, in spite of ourselves, one people." A popular slogan suggested that the effect of the electric telegraph would be to "make muskets into candlesticks." Indeed, the construction of a global telegraph network was widely expected, by Briggs and Maverick among others, to result in world peace: "It is impossible that old prejudices and hostilities should longer exist, while such an instrument has been created for the exchange of thought between all the nations of the earth."

The transatlantic cable was regarded as nothing short of miraculous; indeed, it was a miracle that it worked at all. The cable was so unreliable that it was more than a week before the first message was sent successfully, and it took sixteen and a half hours to send Queen Victoria's message to President Buchanan. The official opening of the cable to public traffic was delayed again and again, and commercial messages started to pile up at both ends, while the true state of affairs was kept under wraps. The reliability of the cable steadily deteriorated, and it eventually stopped working altogether on September 1, less than a month after its completion.

THE NEWS THAT the Atlantic cable had failed caused an outcry, not to mention a great deal of embarrassment. Some even claimed the whole thing had been a hoax—that there had never been a working cable, and it was all an elaborate trick organized by Field to make a fortune on the stock market. "Was the Atlantic cable a humbug?" asked the *Boston Courier*, in an article that suggested that the message from Queen Victoria to President Buchanan had in fact been sent weeks in advance by ordinary mail. In a bid to silence the skeptics, the full transcript of the messages sent over the cable before it failed was released. It makes sorry reading and is all too convincing—most of the messages were along the lines of "CAN YOU RECEIVE ME?" and "PLEASE SAY IF YOU CAN READ THIS," as the operators at each end struggled to get the cable to work. The following year, another high-profile telegraphic venture, an attempt to build a submarine cable through the Red Sea to India funded by the British government, also ended in failure. This time, since it was public money that had been lost, there were widespread calls for a public inquiry.

A joint committee was appointed, with four representatives from the Atlantic Telegraph Company and four chosen by the British government, including Professor Wheatstone. For several months the committee took evidence from witnesses, both expert and not-so-expert, in an attempt to get to the bottom of the problem of long-distance submarine telegraphy. The star witness, and the man who did more than anyone else to put submarine telegraphy on a firm scientific footing, was William Thomson, professor of natural philosophy at Glasgow University—and, by this time, the archrival of the Atlantic cable's designer, Dr. Whitehouse.

Whitehouse had been conveniently ill and unable to go to sea to lay the failed cable, and it was Thomson who had kindly agreed to stand in for him, even though he had grave reservations about its design. He had already done a lot of theoretical work on the nature of submarine cables, and his measured, scientifically justified evidence to the public inquiry made mincemeat out of Whitehouse. Not only had Whitehouse made the conducting core too small, Thomson explained, but his use of high-voltage induction coils had gradually destroyed the cable's insulation and caused its eventual demise.

Worse still, Whitehouse had disobeyed his superiors and had acted as though the sole purpose of the cable's existence was to satisfy his experimental curiosity. When it became clear that a highly sensitive new kind of receiving apparatus, the mirror galvanometer, was better suited for transatlantic telegraphy than his own patented automatic receiver, Whitehouse grudgingly agreed to use it—even though he continued to claim that messages were in fact being received on his apparatus. This did little to endear him to Professor Thomson, who had invented the mirror galvanometer.

Exasperated by Whitehouse's behavior, the directors of the Atlantic Telegraph Company eventually sacked him. He retaliated almost immediately by publishing a book called *The Atlantic Telegraph* in a bid to protect his reputation. It was an extraordinarily one-sided account of the Atlantic cable. As he attempted to defend himself and his flawed theories, Whitehouse blamed everyone around him. Portraying himself as a man of science struggling against the forces of ignorance and incompetence, he blamed the manufacturers of the cable, the crew of the ship that laid it, and, most of all, Cyrus Field and the other officials of the Atlantic Telegraph Company, who he claimed had refused to let him carry out all the tests he recommended. He denounced Thomson's new theory of electricity as "fiction" and derided his mirror galvanometer as impractical. Whitehouse was so convinced that he understood telegraphy better than anyone else that he even invented his own "improved" version of Morse code. He also had what he thought was a brilliant new idea—codebooks containing numbered words, which he evidently didn't realize had been abandoned by both Chappe and Morse years earlier.

Thomson's evidence to the committee, and the public savaging of Whitehouse that took place in the letters pages of the *Engineer* journal, ruined Whitehouse's reputation as effectively as the huge voltages generated by his induction coils had ruined the cable. Conveniently for the Atlantic Telegraph Company, which ought to have taken some of the blame for rushing the manufacture of the cable, the responsibility for the failure could be laid firmly at White-house's door. With him gone, the company argued, the mistakes that had caused the cable to fail would never be repeated. Thomson, meanwhile, had shown that he under-stood the theory of submarine telegraphy; a theory that was vindicated in 1864 when a cable was successfully laid linking India with Europe via the Persian Gulf, along which messages were sent using low voltages and Thom-son's sensitive mirror galvanometer as a detector. This time, it really looked as though the problems of submarine telegraphy had been solved, and Field was soon able to raise the money for a new Atlantic cable.

THE NEW CABLE was built with a lot more care than its predecessor. Following Pro-fessor Thomson's recommendations, it had a much larger conducting core; it was also more buoyant, so it would be less likely to snap under its own weight. Still, it was so heavy that there was only one ship in the world that would be able to carry it: the *Great Eastern*, designed by Isambard Kingdom Brunel, and easily the largest ship afloat. The *Great Eastern* had proved something of a white elephant; its large size should have resulted in huge economies of scale, but mismanagement and bad luck meant it had never made anyone any money. The ship was, however, ideally suited for cable laying, and on June 24, 1865, with the new cable loaded onto three vast drums, it set out for Valentia.

A month later, having laid the Irish end of the cable, the *Great Eastern* headed west across the Atlantic, paying out the cable as it went. The cable was tested regularly, and whenever a fault was found, the cable was cut, the ship turned around, and the cable was hauled back in until the faulty part revealed itself. However, on August 2, two-thirds of the way across the Atlantic, the cable broke during one of these splicing operations and disappeared under the waves, into water two miles deep. Several attempts to recover the cable were made with grapnels and improvised steel wires, but every time it was lifted to the surface, the steel wires broke. Eventually the *Great Eastern* turned around and headed back toward Europe.

Despite this failure, raising the money for a third cable did not prove too difficult; the Atlantic Telegraph Company now had so much experience in cable laying that it seemed certain to succeed. What's more, armed with the right equipment, Field was confident of being able to recover the second cable. The following year, on the apparently inauspicious date of Friday, July 13, the *Great Eastern* set out from Valentia again trailing a new cable from an improved paying-out mechanism. Two weeks later, after an uneventful voyage, it reached Newfoundland, and the cable was secured. Once again, Europe and North America had been linked.

Demand for the new cable was so great that on its first day of operation it earned a staggering £1,000. And within a month the *Great Eastern* had successfully recovered the lost cable of the previous year from two miles down on the seabed. More cable was spliced on, and there were soon two working telegraph links across the Atlantic. The death blow was finally dealt to Whitehouse's high-voltage theories by the noted engineer Josiah Latimer Clark, who had the two cables connected back-to-back and successfully sent a signal around the whole circuit—from Ireland to Newfoundland and back—using a tiny battery and Thomson's mirror galvanometer as the detector. The electric telegraph had finally conquered the Atlantic.

THIS TIME THERE WAS NO question of the cable being a hoax. Thomson was knighted, and Congress gave Field a unanimous vote of thanks and awarded him a specially minted gold medal. Honors were also bestowed on Wheatstone and Cooke, and, belatedly, Francis Ronalds, whose original plans for an electric telegraph had been rejected by the Admiralty half a century earlier. (Thomson went on to become Lord Kelvin, after whom the unit of temperature used by scientists is named.)

The hype soon got going again when it became clear

that, this time, the transatlantic link was here to stay. At a banquet held in Field's honor by the New York Chamber of Commerce in November 1866, he was described as "the Columbus of our time. . . . he has, by his cable, moored the New World close alongside the Old." His life's work, the transatlantic cable, was hailed as "the most wonderful achievement of our civilization."

The cables were so profitable that Field was able to pay off all his debts in 1867. That year, when one of the two cables got crushed by an iceberg and stopped working, it was repaired within weeks. Before long, the recovery and repair of undersea cables was regarded as commonplace.

Another banquet was held for Morse at Delmonico's in New York in December 1868, where he was toasted for having "annihilated both space and time in the transmission of intelligence. The breadth of the Atlantic, with all its waves, is as nothing."

Echoing the sentiments expressed on the completion of the 1858 cable, a toast proposed by Edward Thornton, the British ambassador, emphasized the peacemaking potential of the telegraph. "What can be more likely to effect [peace] than a constant and complete intercourse between all nations and individuals in the world?" he asked. "Steam [power] was the first olive branch offered to us by science. Then came a still more effective olive branch—this wonderful electric telegraph, which enables any man who happens to be within reach of a wire to communicate instantaneously with his fellow men all over the world." And another toast was to "the telegraph wire, the nerve of international life, transmitting knowledge of events, removing causes of misunderstanding, and promoting peace and harmony throughout the world."

Was there no limit to the telegraph's power to amaze? Well, actually, there was. For just as the reach of the telegraph was starting to expand across the oceans, some parts of the network were so congested that its whole raison d'être—the rapid delivery of messages—was being undermined. As the volume of traffic increased, the telegraph was in danger of becoming a victim of its own success.

6.

steam-powered messages

> The demands for the telegraph have been constantly
> increasing; they have been spread over every civilized
> country in the world, and have become, by usage,
> absolutely necessary for the well-being of society.
> —*NEW YORK TIMES*, April 3, 1872

S PEEDY COMMUNICATION is a marvelous
thing. But as anyone who uses e-mail will
testify, once you've got used to being able to send messages
very quickly, it's very difficult to put up with delays. Just
as today's e-mail systems are still plagued by occasional
blackouts and failures, the telegraph networks of the 1850s
were subject to congestion as the volume of traffic mush-
roomed, and key network links within major cities became
overloaded.

The problem arose because most telegraph messages

were not transmitted directly from the telegraph office nearest the sender to the one nearest the recipient, but passed via one or more intermediate points where they were retranscribed and retransmitted each time. At busy times, messages might be coming in to a particular telegraph office faster than that office could handle them. Instead of being immediately retransmitted, the messages, transcribed on slips of paper, literally started to pile up.

Parts of the London network were, in fact, so congested that complaints about delayed messages were soon a common refrain within the business community. A cartoon published in *Punch* magazine in 1863 showed two gentlemen lamenting the sorry state of the telegraph system. "What an age we live in," complains one. "It is now six o'clock, and we are in Fleet Street and this message was only sent from Oxford Circus yesterday afternoon at three." (Fleet Street is less than half an hour's walk from Oxford Circus.) Stories like this threatened to undermine public confidence in the telegraph's legendary speed and efficiency.

Some telegraph companies tried employing additional messenger boys to carry bundles of messages along busy routes from one telegraph station to another—a distance of only a few hundred yards in many cases. With enough messages in a bundle, this method was quicker than retelegraphing them, but it hardly inspired public confidence in the new technology. Instead, it gave the impression that the telegraph system was merely a glorified and far more expensive postal service. On the other hand, because the number of messages being transmitted over busy parts of the network varied so dramatically, simply installing more telegraph lines and staffing them with more operators wasn't practical either; for if there were very few messages to handle during a lull, the highly paid operators would have nothing to do. A cheap, efficient way had to be found to transfer large numbers of messages over those branches of the network that were prone to sudden surges in traffic. Something new was called for—and fast.

In LONDON, the problem of congestion first emerged in the early 1850s, when half of all telegraph messages related to the Stock Exchange, another third were business related, and only one in seven concerned "family affairs." In other words, the main use of the telegraph was to send time-sensitive information between the Stock Exchange and other parts of the country. As a result, the telegraph link between the Stock Exchange branch office and the Central Telegraph Office, a distance of 220 yards, carried more messages than any other part of the network; and the value of these messages depended on their being delivered swiftly.

Josiah Latimer Clark, who worked as an engineer for the Electric Telegraph Company (and who later carried out the experiment that disproved Whitehouse's theories about transatlantic telegraphy), applied himself to the problem and came up with a radical solution. He proposed a steam-powered pneumatic tube system to carry telegraph forms the short distance from the Stock Exchange to the main telegraph office. Since outgoing messages would be carried by tube, the telegraph wire along the route could be dedicated to incoming messages, and the level of traffic along the wire would be dramatically reduced.

Clark first tested the idea in 1853, and by 1854 an airtight tube an inch and a half in diameter had been laid underground between the two telegraph stations. It was capable of carrying up to five messages at once, written on telegraph forms and stuffed into a cylindrical carrier made of the ever-useful gutta-percha. Each carrier had a felt pad at the front end to act as a buffer, and was covered with leather to prevent the gutta-percha from melting, since friction with the inside of the tube tended to make the carriers, moving at twenty feet per second, very hot. A six-horsepower steam engine in the basement of the Central Telegraph Office created a partial vacuum in front of the carrier, and it took about half a minute to draw each one down the tube from the Stock Exchange. Even when the carriers were not fully loaded, this system was much faster than sending the messages by telegraph, which could only send about one message per minute. Once a carrier arrived at the Central Telegraph Office, the forms were unloaded and the messages telegraphed to their destinations in the usual way. The original tube was one-way only, since the vast majority of messages originated at the Stock Exchange end. Batches of empty carriers were taken back to the Stock Exchange by messenger.

This first pneumatic tube was far from perfect, and carriers frequently got stuck, but the company was convinced of the benefits and introduced a second underground tube in 1858. With a larger bore (two and a quarter inches) and running nearly a mile from another branch office, in Mincing Lane, to the central office, this improved tube was operated by a more powerful twenty-horsepower steam engine. It proved successful enough that after a while the company decided to make this tube two-way.

So that steam engines would not be needed at both ends, a "vacuum reservoir," consisting of an airtight, lead-covered box, ten by twelve by fourteen feet, was constructed in the basement of a house in Mincing Lane. However, one day a carrier got stuck in the tube, causing the pressure in the vacuum reservoir to drop, until eventually it imploded with a loud bang, demolishing the wall between it and a nearby house. According to a contemporary report, "At the time the landlord of the house happened to be dining in the next room, and he suddenly found himself, his table, his dinner, and the door, which was wrenched off its hinges, precipitated into the room amongst the debris of the chamber." Following this accident, carriers were sent by pushing them along the tubes with compressed air, rather than drawing them along with a partial vacuum.

By 1865, the increase in traffic had led the Electric

Telegraph Company to extend its London tube network and install tube systems in Liverpool, Birmingham, and Manchester. Similar systems were initiated in Berlin in 1865 and Paris in 1866, and before long there were also pneumatic tube networks in Vienna, Prague, Munich, Rio de Janeiro, Dublin, Rome, Naples, Milan, and Marseilles. One of the most ambitious systems was installed in New York, linking many of the post offices in Manhattan and Brooklyn. This system was large enough to handle small parcels, and on one occasion a cat was even sent from one post office to another along the tubes.

By 1870, three-inch-diameter tubes were the norm, with carriers capable of transporting as many as sixty messages, though they were usually sent holding far fewer. According to statistics compiled in London, one three-inch tube was equivalent to seven telegraph wires and fourteen operators working flat out. Tubes were also good for coping with sudden surges in demand, such as when war fever struck London in July 1870 and the amount of traffic instantly doubled.

However, blockages were a constant problem for all pneumatic tube networks. They were usually cleared by blasting air down the tubes—though really serious blockages meant having to dig up the street. In Paris, the distance to the blockage was sometimes calculated by firing a pistol down the tube and noting the time delay before the sound of the bullet's impact with the carrier. Leaks, on the other hand, were harder to find; the preferred method was to send a carrier on the end of a long string, and note the point at which the rate of take-up of the string slackened.

ALTHOUGH THEY WERE ORIGINALLY intended to move messages from one telegraph office to another, pneumatic tube systems were soon being used to move messages around within major telegraph offices. Each of these offices was a vast information processing center—a hive of activity surrounded by a cat's cradle of telegraph wires, filled with pneumatic tubes, and staffed by hundreds of people whose sole purpose was to receive messages, figure out where to send them, and dispatch them accordingly.

The layout of a major telegraph office was carefully organized to make the flow of information as efficient as possible. Typically, pneumatic tube and telegraph links to offices within the same city would be grouped on one floor of the building, and telegraph wires carrying messages to and from distant towns and cities would be located on another floor. Grouping lines in this way meant that additional instruments and operators could easily be assigned to particularly busy routes when necessary. International connections, if any, were also grouped.

Incoming messages arriving by wire or by tube were taken to sorting tables on each floor and forwarded as appropriate over the building's internal pneumatic tube system for retransmission. In 1875, the Central Telegraph Office in London, for example, housed 450 telegraph instruments on three floors, linked by sixty-eight internal pneumatic tubes. The main office in New York, at 195 Broadway, had pneumatic tubes linking its floors but also employed "check-girls" to deliver messages within its vast operating rooms. Major telegraph offices also had a press-room, a doctor's office, a maintenance workshop, separate male and female dining rooms, a vast collection of batteries in the basement to provide electrical power for the telegraphic instruments, and steam engines to power the pneumatic tubes. Operators working in shifts ensured that the whole system operated around the clock.

Consider, for example, the path of a message from Clerkenwell in London to Birmingham. After being handed in at the Clerkenwell Office, the telegraph form would be forwarded to the Central Telegraph Office by pneumatic tube, where it would arrive on the "Metropolitan" floor handling messages to and from addresses within London. On the sorting table it would be identified as a message requiring retransmission to another city and would be passed by internal pneumatic tube to the "Provincial" floor for transmission to Birmingham by intercity telegraph. Once it had been received and retranscribed in Birmingham, the message would be sent by pneumatic tube to the telegraph office nearest the recipient and then delivered by messenger.

APPROPRIATELY ENOUGH for the nation that pioneered the first telegraphs, the French had their own twist on the use of pneumatic tubes. For of all the tube networks built around the world, the most successful was in Paris, where sending and receiving *pneus* became part of everyday life in the late nineteenth century. Like the pneumatic tube networks in many other major cities, the Paris network was extensive enough that many local messages could be sent from sender to recipient entirely by tube and messenger, without any need for telegraphic transmission. In these cases, the telegraph form that the sender wrote the message on actually ended up in the hands of the recipient—which meant that long messages were just as easy to deliver as short messages.

So, in 1879, a new pricing structure was announced: For messages traveling within the Paris tube network, the price was fixed, no matter how long the message. Faster than the post and cheaper than sending a telegram, this network provided a convenient way to send local messages within Paris, though the service was operated by the state telegraph company and the messages were officially regarded as telegrams.

Messages were written on special forms, which could be purchased, prepaid, in advance. These could then be deposited into small post boxes next to conventional mailboxes, handed in at telegraph counters in post offices, or put into boxes mounted on the backs of trams, which were unloaded when the trams reached the end of the line. Once in the system, messages were sent along the tubes to the office nearest the destination and then delivered by messenger. Each message might have to pass through several sorting stations on the way to its destination; it was date-stamped at each one, so that its route could be determined. (The same is true of today's e-mail messages, whose headers reveal their exact paths across the Internet.) No enclosures were allowed to be included with messages, and any messages that broke this rule were transferred to the conventional postal service and charged at standard postal rates.

The scheme was a great success, and the volume of messages being passed around the network almost doubled in the first year. The network was further extended as a result, and for many years messages were affectionately known as *petits bleux,* after the blue color of the message forms.

B Y T H E e a r l y 1 8 7 0 s, the Victorian Internet had taken shape: A patchwork of telegraph networks, submarine cables, pneumatic tube systems, and messengers combined to deliver messages within hours over a vast area of the globe. New cables were being laid all over the world. Malta had been linked to Alexandria in 1868, and a direct cable was laid from

France to Newfoundland in 1869. Cables reached India, Hong Kong, China, and Japan in 1870; Australia was connected in 1871, and South America in 1874.

In 1844, when Morse had started building the network, there were a few dozen miles of wire and sending a message from, say, London to Bombay and back took ten weeks. But within thirty years there were over 650,000 miles of wire, 30,000 miles of submarine cable, and 20,000 towns and villages were on-line—and messages could be telegraphed from London to Bombay and back in as little as four minutes. "Time itself is telegraphed out of existence," declared the *Daily Telegraph* of London, a newspaper whose very name was chosen to give the impression of rapid, up-to-date delivery of news. The world had shrunk further and faster than it ever had before.

Morse's original telegraph line between Washington and Baltimore had hardly started out as a moneymaker; but the more points there were on the network, the more useful it became. By the late 1860s, the telegraph industry, and the submarine cable business in particular, was booming—and every investor wanted a piece of the action. "There can be no doubt that the most popular outlet now for commercial enterprise is to be found in the construction of submarine lines of telegraph," reported the *Times* of London in 1869. By 1880, there were almost 100,000 miles of undersea telegraph cable.

Improvements in submarine telegraphy made it possible to run telegraph cables directly from Britain to out-

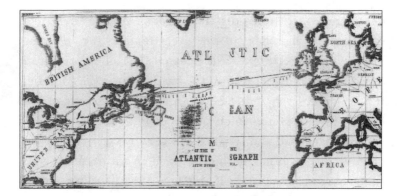

posts of the British Empire, without having to rely on the goodwill of any other countries along the route, and "intra-imperial telegraphy" was seen as an important means of centralizing control in London and protecting imperial traffic from prying eyes. The result was a separate British network that interconnected with the global telegraph network at key points.

As the network connected more and more countries, the peaceful sentiments that had been expressed on the completion of the Atlantic cable were extended to embrace the whole of humanity. The telegraph was increasingly hailed as nothing less than the instrument of world peace.

"It brings the world together. It joins the sundered

hemispheres. It unites distant nations, making them feel that they are members of one great family," wrote Cyrus Field's brother Henry. "An ocean cable is not an iron chain, lying cold and dead in the icy depths of the Atlantic. It is a living, fleshy bond between severed portions of the human family, along which pulses of love and tenderness will run backward and forward forever. By such strong ties does it tend to bind the human race in unity, peace and concord. . . . it seems as if this sea-nymph, rising out of the waves, was born to be the herald of peace."

Or, as another poetically inclined advocate of the telegraph's peacemaking powers put it: "The different nations and races of men will stand, as it were, in the presence of one another. They will know one another better. They will act and react upon each other. They may be moved by common sympathies and swayed by common interests. Thus the electric spark is the true Promethean fire which is to kindle human hearts. Men then will learn that they are brethren, and that it is not less their interest than their duty to cultivate goodwill and peace throughout the earth."

Unfortunately, the social impact of the global telegraph network did not turn out to be so straightforward. Better communication does not necessarily lead to a wider understanding of other points of view; the potential of new technologies to change things for the better is invariably overstated, while the ways in which they will make things worse are usually unforeseen.

7.

CODES, HACKERS, AND CHEATS

Some simple yet secure cipher, easily acquired and easily read, should be introduced, by which means messages might to all intents and purposes be "sealed" to any person except the recipient.

—*QUARTERLY REVIEW*, 1853

ever since people have invented things, other people have found ways to put those things to criminal use. "It is a well-known fact that no other section of the population avail themselves more readily and speedily of the latest triumphs of science than the criminal class," declared Inspector John Bonfield, a Chicago policeman, to the *Chicago Herald* in 1888. "The educated criminal skims the cream from every new invention, if he can make use of it." The telegraph was no exception. It provided unscrupulous individuals with novel opportunities for fraud, theft, and deception.

In the days of the original optical telegraphs, Chappe's suggestion that the network be used to transmit stock market information was originally rejected by Napoleon. However, by the 1830s it was being used for just that purpose, and before long, it was being abused. Two bankers, François and Joseph Blanc, bribed the telegraph operators at a station near Tours, on the Paris-Bordeaux line, to introduce deliberate but recognizable errors into the transmission of stock market information, to indicate whether the stock market in Paris had gone up or down that day. By observing the arms of the telegraph from a safe distance as they made what appeared to everyone else to be nothing more than occasional mistakes, the Blanc brothers could gain advance information about the state of the market without the risk of being seen to be associating with their accomplices. This scheme operated for two years before it was discovered in 1836.

With the telegraph's ability to destroy distance, it provided plenty of scope for exploiting information imbalances: situations where financial advantage can be gained in one place from exclusive ownership of privileged information that is widely known in another place. A classic example is horse racing. The result of a race is known at the racetrack as soon as it is declared, but before the invention of the telegraph, the information could take hours or even days to reach the bookmakers in other parts of the country. Anyone in possession of the results of a horse race before the news reached the bookmakers could then place a surefire bet on the winning horse. Almost immediately, rules were introduced to disallow the transmission of such information by telegraph; but, as is often the case with attempts to regulate new technologies, the criminals tended to be one step ahead of those making the rules.

One story from the 1840s tells of a man who went into the telegraph office at Shoreditch station in London on the day of the Derby, an annual horse race, and explained that he had left his luggage and a shawl in the care of a friend at another station—the station that just happened to be nearest the racetrack. He sent a perfectly innocent-sounding message asking his friend to send the luggage and the shawl down to London on the next train, and the reply came back: "YOUR LUGGAGE AND TARTAN WILL BE SAFE BY THE NEXT TRAIN." The apparently harmless reference to "TARTAN" revealed the colors of the winning horse and enabled the man to place a bet and make a hefty profit.

Another attempt at such a ruse was, however, less successful. A man went into the Shoreditch telegraph office on the day of a major horse race at Doncaster, at around the time of the race's end. He explained that he was expecting an important package on the train from Doncaster, and that a friend had agreed to put the package in one of the first-class carriages. He asked if he could telegraph his friend to find out the number of the carriage. However, the clerk saw through his scheme because the carriages on that particular line—unlike the racehorses in that day's race—were not numbered. According to an account in *Anecdotes of the Telegraph*, when his request was questioned, the man ran off, "grinning a horrible, ghastly smile."

Both of these schemes used what was, in effect, a code, but a cleverly disguised one, since in the early days of the telegraph the use of codes in telegrams was not allowed—except by governments and officials of the telegraph companies.

The Electric Telegraph Company, for instance, sent share prices from London to Edinburgh "by means of certain cabalistic signs"—in other words, in code. Company officials consulted special codebooks to encode the prices in London, and decode them again in Edinburgh, where they were posted on a board in a public room, to which bankers, merchants, and tradesmen could gain access provided they paid a fee. In the early 1840s, before the telegraph network became widespread, this was a good arrangement for all concerned; by transmitting the information over a distance of several hundred miles, the telegraph company was exploiting another information imbalance, turning something that was common knowledge in London into a valuable commodity in Scotland.

Inevitably, an unscrupulous broker, hoping to do some exploiting of his own, tried to get hold of this valuable information without paying for it. He invited two telegraph clerks to a pub and offered them a cut of any profits he made from share price information passed to him. But when he failed to maintain his half of the bargain, they turned against him and exposed him to the authorities.

This classic example shows that no matter how secure a code is, a person is always the weakest link in the chain. Even so, there were continuous efforts to dream up uncrackable codes for use over the telegraph.

Cryptography—tinkering with codes and ciphers—was a common hobby among Victorian gentlemen. Wheatstone and his friend Charles Babbage, who is best known for his failed attempts to build a mechanical computer, were both keen crackers of codes and ciphers—Victorian hackers, in effect. "Deciphering is, in my opinion, one of the most fascinating of arts," Babbage wrote in his autobiography, "and I fear I have wasted upon it more time than it deserves."

He and Wheatstone enjoyed unscrambling messages that appeared in code in newspaper classified advertisements—a popular way for young lovers to communicate, since a newspaper could be brought into a house without arousing suspicion, unlike a letter or a telegram. On one occasion Wheatstone cracked the cipher, or letter substitution code, used by an Oxford student to communicate with his beloved in London. When the student inserted a message suggesting to the young woman that they run away together, Wheatstone inserted a message of his own, also in cipher, advising her against it. The young woman in-

serted a desperate, final message: "DEAR CHARLIE: WRITE
NO MORE. OUR CIPHER IS DISCOVERED!" On another occasion Wheatstone cracked the code of a seven-page letter
written two hundred years earlier by Charles I entirely in
numbers. He also devised a cunning form of encryption,
though it is generally known as Playfair's Cipher after his
friend, Baron Lyon Playfair. Babbage too invented several
ciphers of his own.

And there was certainly a demand for codes and ciphers; telegrams were generally, though unfairly, regarded
as less secure than letters, since you never knew who
might see them as they were transmitted, retransmitted,
and retranscribed on their way from sender to receiver. In
fact, most telegraph clerks were scrupulously honest, but
there was widespread concern over privacy all the same.

"Means should be taken to obviate one great objection—at present felt with respect to sending private communications by telegraph—the violation of all secrecy,"
complained the *Quarterly Review*, a British magazine, in
1853. "For in any case half a dozen people must be cogni-
zant of every word addressed by one person to another.
The clerks of the [telegraph companies] are sworn to secrecy, but we often write things that it would be intolerable
to see strangers read before our eyes. This is a grievous
fault in the telegraph, and it must be remedied by some
means or other." The obvious solution was to use a code.

Meanwhile, the rules determining when codes could
and could not be used were becoming increasingly compli-

cated as national networks, often with different sets of rules, were interconnected. Most European countries, for example, forbade the use of codes except by governments, and in Prussia there was even a rule that copies of all messages had to be kept by the telegraph company. There were also various rules about which languages telegrams could be sent in; any unapproved language was regarded as a code.

The confusion of different rules increased as more countries signed bilateral connection treaties. Finally, in 1864, the French government decided it was time to sort out the regulatory mess. The major countries of Europe were invited to a conference in Paris to agree on a set of rules for international telegraphy. Twenty states sent delegates, and in 1865 the International Telegraph Union was born. The rules banning the use of codes by anyone other than governments were scrapped; at last, people could legally send telegrams in code. Not surprisingly, they started doing so almost immediately.

In the united states, where the telegraph network was controlled by private companies rather than governments, there were no rules banning the use of codes, so they were adopted much earlier. In fact, the first known public codes for the electric telegraph date back to 1845, when two codebooks were published to provide businesses with a means of communicating secretly using the new technology.

That year, Francis O. J. Smith, a congressman and lawyer who was one of Morse's early backers, published *The Secret Corresponding Vocabulary Adapted for Use to Morse's Electro-Magnetic Telegraph*. At around the same time, Henry J. Rogers published a *Telegraph Dictionary Arranged for Secret Correspondence Through Morse's Electro-Magnetic Telegraph*.

Both codes consisted of nothing more than a numbered dictionary of words (A1645, for example, meant "alone" in Smith's vocabulary, which consisted of 50,000 such substitutions), but since numbers were frequently garbled in transmission—telegraph clerks were used to transmitting recognizable words, not meaningless strings of figures—devisers of codes soon switched to the use of code words to signify other words or even entire phrases. By 1854, one in eight telegrams between New York and New Orleans was sent in code. The use of the telegraph to send bad news in an emergency is well illustrated by one code, which used single Latin words to represent various calamities: *COQUARUM* meant "ENGAGEMENT BROKEN OFF," *CAMBITAS* meant "COLLARBONE PUT OUT," and *GNAPHALIO* meant "PLEASE SEND A SUPPLY OF LIGHT CLOTHING."

Of course, such codes weren't all that secret because the codebooks were widely available to everyone (though in some cases they could be customized). But before long another advantage of using such nonsecret codes, known as "commercial" codes, soon became clear—to save money.

By using a code that replaced several words with a single word, telegrams cost less to send.

Those for whom security rather than economy was paramount preferred to use ciphers, which take longer to encode and decode (since individual letters rather than whole words are substituted) but are harder to crack. Yet while such codes and ciphers were a boon for users, they were extremely inconvenient for telegraph companies. Codes reduced their revenue, since fewer words were transmitted, and ciphers made life harder for operators, who found it more difficult to read and transmit gibberish than messages in everyday language.

The increased difficulty of transmitting gibberish was recognized by the International Telegraph Union (ITU), so when new rules were drawn up concerning the use of codes and ciphers, the convention was adopted that messages in code would be treated just like messages in plain text, provided they used pronounceable words to stand for complex phrases, and that no word was more than seven syllables long. Messages in cipher (defined as those consisting of gibberish words), on the other hand, were charged on the basis that every five characters counted as one word. Since the average length of a word in a telegram was more than five letters, this effectively meant that messages in cipher were charged at a higher rate.

During the 1870s, demand for codes was further fueled by the growth in submarine telegraphy, which allowed messages to be sent to distant lands—for a price. The ABC Code, compiled by William Clausen-Thue, a shipping manager, was the first commercial code to sell in really large quantities. It had a vast vocabulary that represented many common phrases using a single word, which was a particular advantage when sending intercontinental telegrams, since they were extremely expensive. (The original rate for transatlantic telegrams was £20—then about $100—for a minimum of ten words. The cable actually became more profitable when the rate was halved and then halved again, because the lower prices attracted more customers.) On long cables where 90 percent of messages were business related, typically 95 percent were in code.

With codes in such common use, many companies chose to develop their own codes for private use with their own correspondents overseas, either because they wanted additional security or because existing codes failed to meet the vocabulary needs of those in specialized fields. Detwiller & Street, a fireworks manufacturer, for example, devised its own code in which the word "FESTIVAL" meant "A CASE OF THREE MAMMOTH TORPEDOES." In India, the Department of Agriculture had a special code that dealt specifically with weather and famine, in which the word "ENVELOPE" meant "GREAT SWARMS OF LOCUSTS HAVE APPEARED AND RAVAGED THE CROPS." The fishing industry, the mining industry, the sausage industry, bankers, railroads, and insurance companies all had their own codes, often running into hundreds of pages, to detail incredibly specific phrases and situations.

With the use of one commercial code, for example, the following lengthy message—"FLOUR MARKET FOR COMMON AND FAIR BRANDS OF WESTERN IS LOWER, WITH MODERATE DEMAND FOR HOME TRADE AND EXPORT; SALES, 8000 BUSHELS. GENESSEE AT $5.12. WHEAT, PRIME IN FAIR DEMAND, MARKET FIRM, COMMON DESCRIPTION DULL, WITH A DOWNWARD TENDENCY; SALES, 4000 BUSHELS AT $1.10. CORN, FOREIGN NEWS UNSETTLED THE MARKET; NO SALES OF IMPORTANCE MADE. THE ONLY SALE MADE WAS 2500 BUSHELS AT 67C"—can be reduced to: "BAD CAME AFT KEEN DARK ACHE LAIN FAULT ADOPT," a mere nine words.

By 1875, the use of commercial codes was starting to get out of hand. Some codes involved some weird words, like "CHINESISKSLUTNINGSDON." Six syllables, sure—but hardly easy to pronounce, and twenty-one letters long. The telegraph companies felt that too many people were bending the rules. So in 1875, the ITU tried to clamp down on this sort of thing by imposing a fifteen-letter limit. Inevitably, the result was a flurry of new codes that conformed to the new rules but used bogus (though shorter) words like "APOGUMNOSOMETHA."

In 1885, the rules were further tightened. A limit was imposed of ten letters per word for telegrams in code language, and words had to be genuine words in German, English, Spanish, French, Italian, Dutch, Portuguese, or Latin.

What's more, the sending office could demand proof that a word was genuine. Again, new codes were immediately devised in response to the new rules. Every move the telegraph companies made to try to reduce the use of codes was neutralized by the increasing cunning of code compilers.

However, by this stage the drawbacks of such codes were becoming apparent to their users as well as the telegraph companies. Each code word meant so much that a single misplaced letter (or dot or dash) in transmission could dramatically change the meaning of a message.

One particularly graphic example occurred in June 1887, when Frank J. Primrose, a wool dealer in Philadelphia, sent William B. Toland to Kansas to act as his agent and buy wool on his behalf. Using a widely available off-the-shelf commercial code, the two men passed several messages back and forth as they kept each other informed of their transactions. But things went horribly wrong when Primrose sent a message explaining that he had bought 500,000 pounds of wool. The words "I HAVE BOUGHT" were encoded by the word "BAY" in the commercial code, and the amount 500,000 pounds by the word "QUO," so that "I HAVE BOUGHT ALL KINDS, 500,000 POUNDS" became "BAY ALL KINDS QUO."

This message was incorrectly transmitted to Toland as "BUY ALL KINDS QUO," possibly because the Morse code for "A" (dot dash) differs by only one dot from the Morse code for "U" (dot dot dash). As a result, Toland assumed he was being instructed to "buy all kinds, 500,000 pounds" and duly started to buy half a million pounds of wool. By the time the mistake had been uncovered, the market had turned and Primrose ended up losing $20,000. He tried to sue Western Union, the telegraph company that had transmitted the fateful message, but he lost because he had failed to ask for the message to be verified—an optional service that would have cost him a few cents extra. Eventually, after a lengthy legal battle, the Supreme Court ruled that he was entitled to a refund only on the cost of sending the original telegram, or just $1.15.

To prevent these sorts of errors, new codes were devised where words specific to a particular industry were excluded from use as code words to avoid confusion, and all remaining words were chosen so that they all differed from each other by at least two letters. That way, if one letter got garbled in transmission, there would be no danger that it would be taken for another code word with a different meaning. Special look-up books were provided to help with error correction.

However, the number of genuine words of ten characters or less that differed by at least two characters from all other words was very small, so once again the code compilers started to bend the rules and use deliberately misspelled words. This was technically not allowed, because code words were supposed to be genuine words in one of the allowed languages, but the code compilers knew that the telegraph clerks could hardly be expected to know how to spell every single word in all of the allowed languages.

But by 1890, the ITU had got wind of this ruse too and decided the only solution was to compile an official vocabulary of all permitted words; any word not in the official vocabulary would then be charged at the cipher rate. In 1894, the first edition of the vocabulary was published; it contained 256,740 words of between five and ten letters, drawn from each of the eight permitted languages. But it was widely criticized—not least because so many common words were omitted. So the ITU scrapped it and decided to try again. Work started on a new vocabulary containing millions of words, but the plan was abandoned when the impracticability of printing thousands of copies of the vast vocabulary, and getting telegraph clerks to laboriously check every word of every message, became apparent.

In other words, as fast as the rules were changed, new codes were devised to get around them. And eventually users got what they wanted—the ability to send coded messages.

A PARTICULARLY important use of codes was by banks. Worries about the security of telegraphic money transfers were holding back the development of on-line commerce ("The opportunity for fraud has been the chief obstacle," declared the *Journal of the Telegraph* in 1872), so banks started to rely on sophisticated private codes to ensure the safe transmission of money. Although there were existing schemes for transferring money, they were insecure and depended on a high level of trust between both parties and telegraph operators at each end. There was clearly a need for a more secure system, which would unlock a whole new market as people in need of money in a hurry turned to the telegraph.

In 1872, Western Union (by then the dominant telegraph company in the United States) decided to implement a new, secure scheme to enable sums of up to $100 to be transferred between several hundred towns by telegraph. The system worked by dividing the company's network into twenty districts, each of which had its own superintendent. A telegram from the sender's office to the district superintendent confirmed that the money had been deposited; the superintendent would then send another telegram to the recipient's office authorizing the payment. Both of these messages used a code based on numbered codebooks. Each telegraph office had one of these books, with pages containing hundreds of words. But the numbers next to these words varied from office to office; only the district superintendent had copies of each office's uniquely numbered book.

A running count was kept for each book, and each time a money transfer telegram was sent, the next word in its unique numerical order was sent as one of the words of the message. Another page in the codebook gave code words for different amounts in dollars. And a special password, known by the superintendent and the operators at each office, also had to be included, sometimes as the first word in the message, and sometimes as the last word. The system was deemed to be secure enough that up to $6,000 could be transferred between fifteen designated major cities "to meet the occasional exigencies of businessmen."

The service soon became very popular, and by 1877 it was being used to transfer nearly $2.5 million annually in 38,669 separate transactions. "This service, which meets a vast demand from parties caught in unexpected conditions of loss or embarrassment, is one of the greatest boons of our modern civilization," wrote James Reid, a chronicler of the industry, in 1878. Even so, the misunderstandings about the exact nature of the telegraph continued. One woman went into a telegraph office to wire the sum of $11.76 to someone and then changed the amount to $12 because she said she was afraid that the loose change "might get lost traveling over the wire."

B UT EVEN WITH the introduction of security measures, there were still ways to make money by abusing the telegraph. In 1886, forty years after the first attempts to use the telegraph to make money on the horses, an Englishman called Myers tried to bribe an operator at the Exchange Telegraph Company's office at the Haymarket in London to delay the transmission of racing results so that he could place bets on the winners. He was arrested, but when the case went to court it was found that the only telegraph-related law he could be charged under related to damaging telegraphic apparatus, something of which he was clearly not guilty. Delaying the mail was illegal, but delaying a telegram was not. The law was subsequently extended to make it a crime to alter, delay, or disclose the contents of a telegram. Myers's case was never tried because he killed himself with an overdose of laudanum. But it highlighted yet another example of technological progress outstripping the lawmakers.

There were, of course, instances where intercepting telegrams was regarded as acceptable—when it was governments doing the intercepting, in the interests of national security. As a result, diplomats and spies routinely used codes and ciphers to protect their messages from the prying eyes of enemy governments, but with varying degrees of success. Perhaps the most notorious example of an intercepted telegram was the Panizzardi telegram, and its unfortunate consequences for Captain Alfred Dreyfus of the War Ministry in Paris—an episode that became known as the Dreyfus Affair.

On October 15, 1894, Captain Dreyfus, an artillery officer, was summoned to a meeting and asked to take some dictation. Once he had written a few words, his writing was compared with a newly discovered document written by a treacherous staff member at the War Ministry who was passing information to the Germans. On the basis of his handwriting, Dreyfus was accused of being the author of the incriminating document and was arrested on the spot for high treason.

Two weeks later the news was leaked, and a newspaper, *La Libre Parole*, reported that Dreyfus had been arrested for spying and was suspected of being in the pay of Germany or Italy. The resulting outcry was to split French society into two factions: Dreyfusards (broadly, liberals who believed Dreyfus had been framed) and anti-Dreyfusards (pro-military conservatives who thought he was guilty). Since Dreyfus was Jewish, the anti-Dreyfusards were accused of anti-Semitism, and the polarization of political views that followed inflamed anti-Jewish feeling and divided the country.

As tension mounted, the Italian military attaché, Colonel Alessandro Panizzardi, sent a telegram to his chief in Rome to say that as far as he knew, Dreyfus was not spying for them, though there was always the possibility that Dreyfus reported directly to someone higher up the chain of command in Rome. With speculation rampant in the press, Panizzardi pressed his superiors in Rome to issue an official statement if indeed Dreyfus was not one of its spies. Panizzardi sent a telegram to this effect to Rome—a coded message that was to become one of the most infamous telegrams ever sent.

The message was sent in a numerical commercial code, where different groups of numbers represented different syllables, letters, and common words. And like all diplomatic telegrams, it was immediately intercepted by the French Ministry of Posts and Telegraphs, and a copy was sent to the Foreign Ministry so that the codebreakers at the Bureau du Chiffre (literally, office of code) could work on it. (The French were, once again, ahead of the field; France was, at the time, the only country to have an official military codebreaking effort.)

The codebreakers soon recognized the numerical groups. They were from a commercial code published a few months earlier by an Italian codemaker, Paulo Baravelli. The code used single digits to represent vowels and punctuation marks, double digits for consonants and certain common verbs, triple digits for common syllables, and quadruple digits for key words. This system made messages written in the Baravelli code easy to spot.

In fact, the codebreakers already knew the Baravelli code from earlier that year, when a torrent of telegrams had been exchanged between the count of Turin, a nephew of the king of Italy, and Duchess Graziolo, a legendary Italian beauty staying in Paris. The head of French army intelligence thought this had all the hallmarks of a spy communicating with his spymaster, so he ordered the messages to be decoded. But nobody could make heads or tails of any of them, because they were all written in numbers. Eventually, a French agent broke into the duchess's rooms and found a small, highly scented book: her Baravelli codebook. The messages were soon decoded and were found to express nothing more than what one official de-

scribed as "simple, elemental, natural feelings"—a correspondence between lovers, not spies. The Baravelli code
thus became known to the Bureau du Chiffre.

However, like many commercial codes, the Baravelli
code could be customized, to ensure that messages encoded with it could not simply be read by anyone who had
a copy. Each page contained one hundred words numbered
from oo to 99. These were made into four-digit groups by
combining them with the number of the page on which
they appeared. But each page also had a blank where an
alternative page number could be filled in. By renumber-
ing the pages of two Baravelli codebooks in an identical
fashion, two people could then exchange messages with a
fair amount of secrecy—since there are an astronomical
number of ways in which one hundred pages can be reor-
dered. What's more, some pages had blanks where additional words could be filled in, words whose meaning
would be unknown to anyone who intercepted any message.

When the codebreakers tried to read Panizzardi's telegram using the original page numbers, it was gibberish; he
was clearly using his own page numbers for added security.
However, since one word of the message—"DREYFUS"—was
known, it wasn't all that hard to figure out some of the
page number substitutions, and eventually a partially decoded message was produced: "IF CAPTAIN DREYFUS HAD
NOT HAD RELATIONS WITH YOU, IT WOULD BE WISE FOR THE
AMBASSADOR TO DENY IT OFFICIALLY." Only the meaning of

the end of the message was uncertain; the codebreakers' best guess was "OUR EMISSARY IS WARNED."

This ambiguous message was taken by the French chief of staff, an anti-Dreyfusard who was anxious to secure a conviction, as proof that Dreyfus was guilty. So when, a few days later, the codebreakers decided that the last portion of the message actually meant "to avoid press comment," their superiors were not pleased. There was only one way to see which interpretation was correct: to get Panizzardi to send another message, whose exact wording was known, in the same code. A double agent duly passed a bogus morsel of information to Panizzardi, and when he passed it on in coded form to his superiors in Rome, a copy was passed back to the Bureau du Chiffre. When they decoded it, it confirmed their second interpretation; Dreyfus was innocent.

However, by this time the army was unwilling to admit that it had got the wrong man, so an exaggerated version of the original flawed decoding was presented at Dreyfus's trial. (The complexity of the code meant that the telegram could be claimed to have said almost anything.) Consequently, Dreyfus was found guilty and sent to prison on Devil's Island, a colony off the coast of French Guiana.

Perhaps appropriately for someone wrongly convicted by a telegram, Dreyfus was eventually freed as a result of another telegram. In 1896, the contents of a wastepaper basket from the office of the German military attaché in Paris were examined by French intelligence staff and found to include a torn-up pneumatic telegram form that had never been sent. When pieced together, it was found that the form contained a message to another officer at the French War Ministry, Major Ferdinand Walsin Esterhazy, implicating him in the offenses attributed to Dreyfus. It was, however, another ten years before Dreyfus was finally reinstated. By this time, his case had become a cause célèbre among Paris intellectuals led by the novelist Émile Zola, who wrote his famous article "J'Accuse" in Dreyfus's defense. (The case remained such a political hot potato that it was only in 1995 that the French army finally conceded that Dreyfus had been innocent all along.)

So much for universal peace and understanding. The telegraph was providing new ways to cheat, steal, lie, and deceive.

8.

Love over
the wires

All the ends of the earth will be wooed into
the electric telegraph circuit.

—*SCIENTIFIC AMERICAN*, 1852

SPIES AND CRIMINALS are invariably among the first to take advantage of new modes of communication. But lovers are never far behind.

There are no known examples of amorous messages being passed over the optical telegraph, since it was not available for general public use; but within a few months of the electric telegraph being opened to the public, it was being used for something that even the most farsighted of telegraph advocates had never dared to imagine: to conduct an on-line wedding.

The bride was in Boston, and the groom in New York; the exact date is unknown, but the story of the wedding was common currency by the time a small book, *Anecdotes of the Telegraph*, was published in London in 1848. It was described as "a story which throws into the shade all the feats that have been performed by our British telegraph."

The daughter of a wealthy Boston merchant had fallen in love with Mr. B., a clerk in her father's countinghouse. Although her father had promised her hand to someone else, she decided to disregard his intentions and marry Mr. B. instead. When her father found out, he put the young man on a ship and sent him away on business to England.

The ship made a stopover in New York, where the young woman sent her intended a message, asking him to present himself at the telegraph office with a magistrate at an agreed-upon time. At the appointed hour she was at the other end of the wire in the Boston telegraph office, and, with the telegraph operators relaying their words to and fro in Morse code, the two were duly wed by the magistrate. "The exchange of consent being given by the electric flash, they were thus married by telegraph," reports a contemporary account.

Surprisingly, the marriage was deemed to be legally binding. The woman's father tried to insist that his daughter marry the man he had chosen for her, but she said she was already married to Mr. B., who was by then on his way to England. The merchant apparently "threatened to protest against the validity of the marriage, but did not carry his threat into execution."

In Britain, meanwhile, there was concern that the telegraph could impede the course of true love. Plans had been put forward to extend the telegraph network along the Caledonian railway to Gretna Green, a village just north of the Scottish border that was renowned as a wedding destination for runaway couples from south of the border, since "marriages of declaration" without a minister or magistrate were legal in Scotland but not in England. The extension of the telegraph would mean that runaway couples traveling by train could no longer outrun the news of their elopement, and a disapproving parent could alert the authorities at Gretna Green before they had even arrived. "What an enemy science is to romance and love!" declared one critic.

Clearly, the new technologies would take some getting used to and offered both advantages and drawbacks to amorous correspondents. In general, though, the high cost and lack of secrecy meant that for the general public, the telegram did not represent much of a threat to traditional letter writing. However, the impact of the telegraph on the love lives of its operators—who spent their working lives communicating with each other over the wires—was far more profound.

TELEGRAPH OPERATORS were members of a closed, exclusive community. They had their own customs and vocabulary, and a strict pecking order, based on the speed at which they could send and receive messages. The finest operators worked in the central telegraph offices of major cities; rural outposts, which handled only a few messages a day, were usually operated on a part-time basis by less experienced operators. But collectively, the world's telegraphers represented an online community encompassing thousands of people, very few of whom ever met face-to-face. And despite the apparently impersonal nature of communicating by wire, it was in fact an extremely subtle and intimate means of communication.

Experienced operators could even recognize their friends merely from the style of their Morse code—something that was, apparently, as recognizable as an individual human voice. Each operator on a particular telegraph line also had a two-letter signature, or "sig," with which to identify themselves on-line; a Detroit operator named Mills, for example, used the signature MS, while Miss A. Edwards, another operator, was known as AE.

Operators often developed partnerships with other operators in distant offices. Thomas Edison described such a pairing, established in the 1860s: "When on the New York No. 1 wire that I worked in Boston, there was an operator named Jerry Borst at the other end. He was a first-class receiver and a rapid sender. We made up a scheme to hold this wire, so he changed one letter of the alphabet and I soon got used to it; and finally we changed

A typical midsize U.S. telegraph office. Operators were seated at wooden tables, each equipped with a Morse key and sounder.

three letters. If any operator tried to receive from Borst, he couldn't do it, so Borst and I always worked together."

Operators often claimed ownership over particular wires in this way; although unofficial, the practice did in fact make for smoother, less error-prone transmission, since the sender and receiver knew each other's capabilities. Even so, since it often resulted in operators sitting by relatively quiet wires in order to stay in contact with their friends while other wires were overloaded, companies tried to prevent it from happening by moving operators around (something that the operators referred to as being "snatched"). One notice, posted in a telegraph office in

1898, read as follows: "Those operators who object to being 'snatched,' as they term it, are reminded that upon entering the employ of this company they were not engaged to work any particular wire or wires, to sit at dead or comparatively idle ones when their services could be utilized elsewhere, and the sooner they realize this the better it will be for all concerned."

During quiet periods, however, the on-line interaction really got going, with stories, jokes, and local gossip circulated over the wires. According to one account, "stories are told, opinions exchanged, and laughs enjoyed, just as if the participants were sitting together at a club." In some cases the tales passing over the wires would find their way into the local newspapers. Most did not because, according to Edison, they were far too smutty or anatomically explicit.

Bored and lonely operators would also play checkers over the wires, using a numbering system to identify the squares of the board that dated back to the chess games played over the Washington-Baltimore line in the 1840s. Some telegraphers in remote outposts even preferred on-line contact with other telegraphers to socializing with the locals. Thomas Stevens, a British telegraph operator stationed in Persia, shunned the local community in favor of telegraphic interaction with other Britons. "How companionable it was, that bit of civilization in a barbarous country," he wrote of his telegraphic friends, some of whom were thousands of miles away. (So much for the telegraph's ability to forge links between people of different nations.)

On one occasion the employees of the American Telegraph Company lines between Boston and Calais, Maine, held a meeting by telegraph after hours. The meeting was attended by hundreds of operators in thirty-three offices along the 700-mile line. Each speaker tapped out his words in Morse code so that "all the offices upon the line received his remarks at the same moment, thus annihilating space and time, and bringing together the different parties, in effect, as near to each other as though they were in the same room, although actually separated by hundreds of miles," according to one account. After passing various resolutions, the employees adjourned the meeting "in great harmony and kindly feeling" after about an hour. (In Britain, *Punch* magazine then suggested that holding parliamentary proceedings by telegraph might restrain some of the more verbose speakers in the house from going on for too long.)

THE TELEGRAPHIC community included a large number of women. By the 1870s, the ratio of men to women at the Western Union main office in New York was two to one; in fact, women telegraphers dated back to 1846, when one Sarah G. Bagley was appointed as an operator in Lowell, Massachusetts, after the opening of the New York–Boston line.

In Britain, female telegraphers were usually the daughters of clergymen, tradesmen, and government clerks, and were typically between eighteen and thirty years old and unmarried. Women were regarded as "admirable manipulators of instruments" well suited to telegraphy (since it wasn't too strenuous), and they could spend the quiet periods reading or knitting. The hours were long, though; most operators, including the women, worked ten hours a day, six days a week.

"Ordinarily an operator can tell a woman the moment he hears her working the wire," claimed the *Western Electrician* magazine in 1891. "He tells by her touch on the key. Women, as a rule, telegraphers say, do not touch the key of their instruments as firmly as men do. Occasionally, however, there is one without this characteristic distinction in style." In most cases, female operators were segregated from the men, and some companies employed a "matron" to keep an eye on them. But while the female operators were physically isolated from their male counterparts, they were of course in direct contact with them over the telegraph network throughout the working day. So it hardly needs saying that many working relationships flowered into on-line romances. According to one writer, "Sometimes these flourished; sometimes they came to an abrupt halt when the operators met for the first time."

"Romances of the Telegraph," an article published in *Western Electrician* in 1891, tells the story of a "pretty little romance" that took place at a remote station out in the desert at Yuma, Arizona, near the border with Mexico. "A more uninviting place of abode could scarcely be found. The station consists merely of a big water tank, a rough shed called the telegraph office, and another shed in which half a dozen trackmen live. During the summer months life at the little desert station was almost unbearable." There was nothing to do, it was unbearably hot and very difficult to sleep, so unsurprisingly the operator at the station, John Stansbury, turned to the telegraph wire for companionship.

An acquaintance soon sprung up between Stansbury and the operator in Banning, California, known as "Mat," whom Stansbury described as a "jolly, cheerful sort of fellow." They soon became firm friends and agreed to spend their vacation together in the mountains hunting and fishing. Every detail of the trip was arranged, with Mat insisting that they take rubber boots for fishing, even though Stansbury said he was quite happy in his bare feet. But at the last minute Mat pulled out of the trip, having decided to take the train to vacation in New Mexico instead, a trip that involved passing through Yuma, Stansbury's station. But by the time Mat arrived at Yuma, Stansbury had been taken ill with a fever and was quite delirious.

"During the days of my agony I was vaguely aware of gentle, womanly hands and a kindly female presence in my sick-room," Stansbury later wrote. "And when I returned to the conscious world I was not surprised to find a fair and pleasant face beside me. Its owner said that she had been on the train when I was found stricken down, and had stayed to minister to my sore need. The idea may seem preposterous, but I believe the foundation for my affection had been laid while the unconsciousness of fever was still upon me, and the affection grew into the deepest love as she cared for me during the days of my convalescence. After a time I ventured to tell her of my love, and to ask her if she would be mine; but I was not prepared for her answer. 'John,' she said, 'do you really mean that you wish to marry a girl that insists upon wearing rubber boots?'

" 'Mat!' I said, for I was completely beaten. Then it flashed upon me. She was the operator at Banning, and I, like a fool, had always taken it for granted that she was a man. I am not going to tell you how I convinced her that I wanted to marry her, boots, and all, but I did it, and here we are on our wedding journey. The Southern Pacific Telegraph Company has lost an operator, but I calculate that I am ahead on the deal."

Minnie Swan Mitchell, a young operator in the 1880s, recalled that "many a telegraph romance begun 'over the wire' culminated in marriage." Ella Cheever Thayer's 1879 novel *Wired Love* even built its plot around an on-line courtship.

Inevitably, some ill-starred on-line romances had repercussions in the real world. One such cautionary tale was described in "The Dangers of Wired Love," an article published in *Electrical World* in 1886. It concerned George W. McCutcheon of Brooklyn, who ran a newsstand assisted by his twenty-year-old daughter, Maggie. Business was booming, so he decided to install a telegraph line, with Maggie as its operator. But he soon discovered that she was "keeping up a flirtation" with several young men over the wires, including Frank Frisbie, a married man who worked in the telegraph office of the Long Island Railroad. Over the wires, Maggie invited Frisbie to visit her, and he accepted. When her father found out, he forbade the visit. Maggie began seeing Frisbie on the sly. McCutcheon tried moving his newsstand, but Maggie soon found work in a nearby telegraph office and resumed the relationship. Eventually, her father pursued her to a rendezvous and threatened to "blow her brains out." She had him arrested, and he was charged with threatening behavior.

On the other hand, the telegraph was sometimes able to help couples transcend real-world barriers. In 1876 William Storey, the operator at an army base at Camp Grant, Arizona, thought he was going to have to call off his wedding; he was unable to obtain a leave of absence to travel to San Diego to get married, and there was no minister at the camp to perform the wedding, so there was little point in his fiancée making the trip to Camp Grant. But then Storey had an idea. "A contract by telegraph is binding, then why," he thought, "can we not be married by telegraph?" He hit upon the plan of inviting his fiancée, Clara Choate, to Camp Grant and asking a minister to perform the marriage ceremony remotely, by telegraph. Choate made the wagon trip to the camp, and the Reverend Jonathan Mann agreed to marry them from 650 miles away in San Diego.

Lieutenant Philip Reade, who was in charge of the telegraph lines in California and Arizona, arranged for the line to be cleared after working hours so that the wedding could take place. He sent out a message to all station managers along the line, informing them that the line would be used "for the purpose of conducting a marriage ceremony by telegraph between San Diego and Camp Grant, and you and your friends are specially invited to be present on the occasion, to assist if necessary and to see that good order is maintained." The operators duly accepted the invitation, and at 8:30 P.M. on April 24 a message arrived from the father of the bride in San Diego that he and the minister were ready to proceed. The minister then read out the marriage service, which was relayed to Camp Grant as he spoke. At the appropriate point in the service the bride and groom tapped on the telegraph key to indicate a solemn "I do." Once the service was over, messages of congratulations flooded in from all the stations on the line. The groom was for many years afterward greeted by fellow telegraph operators, who, upon hearing his name, exclaimed that they had been present at his wedding.

The telegraph could also bring together operators working in the same office. "Making Love by Telegraph," another article published in *Western Electrician* in 1891, tells the story of a particularly difficult wire in the New York telegraph station that was connected to a number of district railway offices manned by inexperienced and often incompetent operators. As a result, most operators rapidly lost their temper whenever they tried to work the wire: "It did not matter how good an operator was or how hard he tried to get along, his patience could not stand the strain for more than a few minutes." But there was one female operator who never had any trouble, so she ended up working the wire. One summer a new operator arrived and was assigned to cover the wire during her lunch hour. He was a good-tempered man, but within ten minutes of taking his station he was involved in "a red hot row" with one of the rural operators. It lasted until the young woman got back, when she graciously straightened matters out. The same thing happened day after day, and after a while the young man started to fall in love with the woman. He realized that "nothing short of an angel could work that wire, and he cultivated her acquaintance. They have been married a long time now," the article concludes, "and he has told the secret to his friends matrimonially inclined, and a number of them have been watching that wire ever since. No young woman has remained there for any length of time, and the watchers all know what it means."

THERE WAS ALSO a dark side to telegraphic interaction; the best operators often felt nothing but scorn toward the small-town, part-time operators they often encountered on-line, who were known as "plugs" or "hams." Speed was valued above all else; the fastest operators were known as bonus men, because a bonus was offered to operators who could exceed the normal quota for sending and receiving messages. So-called first-class operators could handle about sixty messages an hour—a rate of twenty-five to thirty words per minute—but the bonus men could handle even more without a loss in accuracy, sometimes reaching speeds of forty words per minute or more.

Wandering workers who went from job to job were known as "boomers." There were no formal job interviews; applicants were simply sat down on a busy wire to see if they could handle it. Since they could find work almost anywhere, many boomers had an itinerant lifestyle; a great number of them suffered from alcoholism or mental health disorders. In a sense, the telegraph community was a meritocracy—it didn't matter who you were as long as you could send and receive messages quickly—which was one of the reasons that women and children were readily admitted to the profession.

New operators usually started out by filling in on an occasional basis or taking seasonal jobs at parks, summer camps, and resorts, and the more talented would soon gravitate toward the cities. Once a young operator had gained a foothold in a city office, he or she could expect to be subjected to a humiliating induction ritual, known as "salting." Sometimes the operator would be sent bogus messages addressed to "L. E. Phant" or "Lynn C. Doyle." But usually the unwary beginner would be asked to operate a wire with a particularly fast sender at the other end, who would start sending at a reasonable rate but then gradually pick up the pace. As the novice operator struggled to keep up, the other operators in the office would soon gather round to watch, and eventually the operator would be forced to admit defeat and "break." Salting was also known as hazing or rushing.

Young Thomas Edison was legendary for being able to take down messages as fast as anyone could transmit them. Edison was taught Morse code as a teenager by a railway stationmaster, whose three-year-old son he had plucked from the path of an oncoming train. He rapidly became an expert operator, and there are numerous tales of his prowess. At one stage the disheveled Edison took a job in Boston, where the operators thought rather highly of themselves and liked to dress like gentlemen. Taking him for a country bumpkin, they asked a very fast operator in another office to salt him. But as the speed of transmission increased, Edison kept on receiving happily at twenty-five, thirty, even thirty-five words per minute. Finally, having received all the messages without any difficulty, Edison tapped back to his opponent: "Why don't you use your other foot?"

Edison's ability arose from his partial deafness, which meant that he was not distracted by background noise as he sat listening to the clicks of the telegraph receiver. In

Thomas Edison, inventor, telegraph pioneer and operator.

later life, he even turned his deafness to his advantage when courting his second wife, Mina. "Even in my courtship my deafness was a help," he wrote in his diary. "In the first place it excused me for getting quite a little nearer to her than I would have dared to if I hadn't had to be quite close in order to hear what she said. My later courtship was carried on by telegraph. I taught the lady of my heart the Morse code, and when she could both send and receive we got along much better than we could have with spoken words by tapping out our remarks to one another on our hands. Presently I asked her thus, in Morse code, if she would marry me. The word 'Yes' is an easy one to send by telegraphic signals, and she sent it. If she had been obliged to speak it, she might have found it harder."

Edison was, like many telegraphers, an inveterate tinkerer who liked to try out new and improved forms of telegraphic apparatus. He preferred to take the night shift so he could spend the day experimenting in a back room at the telegraph office, and he lived on a frugal diet of apple pie washed down with vast amounts of coffee. But when his experiments went wrong, which they often did, he would lose his job and would have to move on to another town. On one occasion he blew up a telegraph office when mixing new battery acid to a recipe of his own devising; another time he spilled sulfuric acid on the floor of a telegraph office above a bank, and the acid ate through the floor and ruined the carpet and furniture in the office below.

But Edison's prowess as an operator enabled him to progress through the ranks of the telegraphic community, and eventually he went right to the top, working as an engineer and inventor and reporting directly to the directors of the major telegraph companies. Indeed, despite the strange customs and the often curious lifestyle of many operators, telegraphy was regarded as an attractive profession, offering the hope of rapid social advancement and fueling the expansion of the middle class. Courses, books, and pamphlets teaching Morse code to beginners flourished. For the ambitious, it provided an escape route from small towns to the big cities, and for those who liked to move around, it meant guaranteed work wherever they went.

Admittedly, there was a rapid turnover of employees in major offices, and telegraphers often had to endure unsociable hours, long shifts, and stressful and unpleasant working conditions. But to become a telegrapher was to join a vast on-line community—and to seek a place among the thousands of men and women united via the worldwide web of wires that trussed up the entire planet.

9.

war and peace in the global village

All the inhabitants of the earth would be brought into
one intellectual neighborhood.
   —Alonzo Jackman, advocating an Atlantic telegraph
    in 1846

DESPITE THE WIDELY expressed optimism that the telegraphs would unite humanity, it was in fact only the telegraph operators who
were able to communicate with each other directly. But
thanks to the telegraph, the general public became participants in a continually unfolding global drama, courtesy of
their newspapers, which were suddenly able to report on
events on the other side of the world within hours of their
occurrence. The result was a dramatic change in world-
view; but to appreciate the extent to which the telegraph
caused an earthquake in the newspaper business requires

an understanding of how newspapers were run in the pre-telegraph era.

At the beginning of the nineteenth century, newspapers tended to cover a small locality, and news traveled as the papers themselves were carried from one place to another. One journalist, Charles Congdon, complained that in those days there was hardly anything in his local New England newspaper. "In that time of small things," he wrote in his memoirs, "subscribers must have been easily satisfied. The news from Europe, when there was any, was usually about six weeks old, or even older." There were very few letters from foreign correspondents, he noted, which was a good thing, "for most of them were far from interesting."

Today, the common perception of a journalist is someone who will stop at nothing to get hold of a story and rush it into the newsroom. But in the early nineteenth century, newspapers traded on their local coverage, not the timeliness of their news. Congdon tells of one editor who refused a journalist's request to visit a nearby town to report on a speech, saying "Somebody will send us in something about it in two or three days." Some newspapers printed on a different day each week to fit in with the social life of the editor; others rationed the amount of news they printed in a busy week, in case there was a shortage of news the following week. And apart from local stories, most other news was taken from the pages of other papers, which were delivered by post—days after publication. Newspapers reprinted each other's stories freely; news moved so slowly that there was no danger that one paper would steal another's story and be on sale at the same time. The free exchange of information that resulted was beneficial to all concerned, though it meant that news was often days or weeks old by the time it reached its readers.

In addition, some of the larger newspapers had correspondents in foreign countries, who would write in to report the latest news from distant cities. Their letters took weeks to arrive, but before the establishment of the telegraph network, there was no other way to send news. It was commonplace for foreign news to be weeks or months old by the time it appeared in print. The *Times* of London had a particularly extensive network of foreign correspondents, so that its largely business readership could be kept informed of overseas political developments that might affect trade. Foreign reports also reported the arrival and departure of ships and detailed their cargoes. But since the news traveled no faster than the ships that carried it, the January 9, 1845, edition of the *Times* included reports from Cape Town that were eight weeks old and news from Rio that was six weeks old. The delay for news from New York was four weeks, and for news from Berlin a week. And the *Times* was a newspaper that prided itself on getting the news by the fastest means possible.

The existence of a newspaper tax in Britain kept prices artificially high, so the *Times* had the market to itself. But in New York things started to heat up in the 1820s with the fierce competition between the *Journal of Commerce* and the rival *Courier and Enquirer*. Both papers were aimed at business readers and fought to distinguish themselves by being first with the news. They established rival pony expresses between New York and Washington to get the political news sooner, and used fast boats to meet incoming vessels from Europe and get the latest news before they docked. Then, in the 1830s, newspapers became a popular medium with the establishment of cheap, mass-market titles. The ensuing rivalry between the newspapers of the so-called penny press led to an increase in the use of carrier pigeons and ships. One editor, James Gordon Bennett of the *New York Herald*, even agreed to pay one of his sources a $500 bonus for every hour European news arrived at the *Herald* in advance of its competitors. Get the news first, and you'll sell more papers: Increasingly, news was worth money.

So it was clear that the establishment of telegraph lines in the 1840s would change everything. In fact, the second message sent on Morse's Washington-Baltimore line—immediately after "WHAT HATH GOD WROUGHT"—was "HAVE YOU ANY NEWS?" But far from welcoming the telegraph, many newspapers feared it.

ALTHOUGH receiving news by telegraph would seem to be the logical next step from using horses, carrier pigeons, and so on, it was instead viewed as an ominous development. The telegraph could deliver news almost instantly, so the competition to see who could get the news first was, in effect, over. The winner would no longer be one of the newspapers; it would be the telegraph. James Gordon Bennett was one of many who assumed that the telegraph would actually put newspapers out of business; because it put all newspapers on a level playing field, his shenanigans to get hold of the news earlier than his rivals would no longer be an advantage. "The telegraph may not affect magazine literature," he suggested, "but the mere newspapers must submit to destiny, and go out of existence." The only role left for printed publications, it seemed, would be to comment on the news and provide analysis.

Of course, this perception turned out to be wrong. While the telegraph was a very efficient means of delivering news to newspaper offices, it was not suitable for distributing the news to large numbers of readers. And although the telegraph did indeed dramatically alter the balance of power between providers and publishers of information, the newspaper proprietors soon realized that, far from putting them out of business, it actually offered great opportunities. For example, breaking news could be reported as it happened, in installments—increasing the suspense and boosting sales. If there were four developments to a major story during the day, newspapers could put out four editions—and some people would buy all four.

But with news available instantly from distant places,

the question arose, Who ought to be doing the reporting? Reporters as we know them today did not exist. So who should get the news? One suggestion was that telegraph operators, who were sprinkled all over the world, should act as reporters. But the handful of telegraph companies that tried to press operators into journalistic service, and then sell their reports to newspapers, found that operators tended to make pretty hopeless journalists. On the other hand, if each newspaper sent its own writer to cover a far-off story, they all ended up sending similar dispatches back from the same place along the same telegraph wire, at great expense.

The logical solution was for newspapers to form groups and cooperate, establishing networks of reporters whose dispatches would be telegraphed back to a central office and then made available to all member newspapers. This would give newspapers the advantage of a far greater reach than they would have had otherwise, without the expense of maintaining dozens of their own reporters in far-flung places. In the United States, the first and one of the best known of these organizations, which came to be called news agencies, was the New York Associated Press, a syndicate of New York newspapers set up in 1848 that quickly established cozy relationships with the telegraph companies and was soon able to dominate the business of selling news to newspapers.

In Europe, meanwhile, Paul Julius von Reuter was also establishing a news agency. Born in Germany, Reuter

started out working for a translation house that took stories from various European newspapers, translated them
into different languages, and redistributed them. Reuter
soon realized that some stories were more valuable than
others, and that businessmen in particular were willing to
pay for timely information, so he set up his own operation,
using carrier pigeons to supply business information several hours before it could be delivered by mail. Initially
operating between Aix-la-Chapelle and Brussels, the Reuter network of correspondents extended across Europe
during the 1840s. Each day, after the afternoon close of
the stock markets, Reuter's representative in each town
would take the latest prices of bonds, stocks, and shares,
copy them onto tissue paper, and place them in a silken
bag, which was taken by homing pigeon to Reuter's head-
quarters. For security, three pigeons were sent, carrying
copies of each message. Reuter then compiled summaries
and delivered them to his subscribers, and he was soon
supplying rudimentary news reports too.

When the electric telegraph was established between
Aix-la-Chapelle and Berlin, Reuter started to use it alongside his pigeons; when England and France were linked by
telegraph in 1851, Reuter moved to London. His policy was
"follow the cable," so London was the place to be, since it
was both the financial capital of the world and the center
of the rapidly expanding international telegraph network.

Although Reuter's reports of foreign events were initially very business oriented—the only angle his business

customers were interested in was how trade would be affected—he soon started trying to sell his dispatches to newspapers. With the abolition of the newspaper tax in Britain in 1855, several new newspapers sprung up, but only the *Times* was capable of covering foreign news, thanks to its well-established network of correspondents who, after some reluctance, started using the telegraphs. The *Times* preferred to use its own reports rather than buy them from Reuter, and turned down the opportunity to do a deal with him three times. Eventually Reuter proved the value of his service in 1859 when he obtained a copy of a crucial French speech concerning relations with Austria and was able to provide it to the *Times* in London within two hours of its being delivered in Paris. During the ensuing war, with the French and Sardinians on one side, and the Austrians on the other, Reuter's correspondents reported from all three camps—and on one occasion dispatched three separate reports of the same battle from the point of view of each of the armies involved. Even so, the *Times* still preferred to rely on its own correspondents, but Reuter was able to sell his dispatches to its rival London newspapers, thus helping them compete with the *Times* without having their own foreign correspondents.

And readers just couldn't get enough foreign news—the more foreign, the better. Instead of limiting their coverage to a small locality, newspapers were able for the first time to give at least the illusion of global coverage, providing a summary of all the significant events of the day, from all over the world, in a single edition. We take this for granted today, but at the time the idea of being able to keep up with world affairs, and feel part of an extended global community, was extraordinary.

It was great for sales, too. "To the press the electric telegraph is an invention of immense value," declared one journalist. "It gives you the news before the circumstances have had time to alter. The press is enabled to lay it fresh before the reader like a steak hot from the gridiron, instead of being cooled and rendered flavourless by a slow journey from a distant kitchen. A battle is fought three thousand miles away, and we have the particulars while they are taking the wounded to the hospital."

The thirst for foreign news was such that when the first transatlantic telegraph cable was completed in 1858, one of the few messages to be successfully transmitted was the news from Europe, as provided by Reuter. "PRAY GIVE US SOME NEWS FOR NEW YORK, THEY ARE MAD FOR NEWS," came the request down the cable from Newfoundland. And so on August 27, 1858, the news headlines were as follows: "EMPEROR OF FRANCE RETURNED TO PARIS. KING OF PRUSSIA TOO ILL TO VISIT QUEEN VICTORIA. SETTLEMENT OF CHINESE QUESTION. GWALIOR INSURGENT ARMY BROKEN UP. ALL INDIA BECOMING TRANQUIL."

This last headline indicated that the Indian Mutiny, a serious rebellion against British rule that had broken out the year before, had been suppressed. However, General Trollope, commander of the British forces in Halifax, Nova

Scotia, had received an order a few weeks earlier by sea from his superiors in London, asking him to send two regiments of troops back across the Atlantic so that they could be redeployed in India. It is not clear whether or not Trollope was aware of the Reuter report, but it clearly indicated that his troops were no longer needed. A telegram from London to General Trollope countermanding the original order was hurriedly sent down the new Atlantic cable, telling him to stay put, and saving the British government £50,000 at a stroke—more than paying back its investment in the cable. It was one of the last messages to reach North America via the ill-fated cable, which stopped working the next day.

But what if the cable had failed earlier? Had he been aware of the Reuter report, Trollope would have known that there was in fact no need for him to send his troops to India, though he would have no doubt followed orders and sent them anyway. This was just one example of how the rapid and widespread distribution of foreign news had unforeseen military and diplomatic implications—something that was brought home to everyone during the Crimean War.

DURING WARTIME, the existence of an international telegraph network meant that news that had hitherto been safe to reveal to newspapers suddenly became highly sensitive, since it could be immediately telegraphed directly into the hands of the enemy. For years it had been customary in Britain for news of departing ships to be reported as they headed off to foreign conflicts; after all, the news could travel no faster than the ships themselves. But the telegraph meant that whatever information was made available in one country was soon known overseas. This took a lot of getting used to, both by governments and news organizations.

As troops departed for the Crimean peninsula following the declaration of war on Russia by France and Britain in March 1854, the War Ministry in London issued precise details of the number and nature of the forces being deployed. This information was faithfully reproduced in the *Times*, which wanted to capitalize on enthusiasm for the war by providing its readers with as much information as possible. Normally the troops would have outstripped the news of their arrival. But with the telegraph network reaching across Europe to the enemy in St. Petersburg, daily reports of the British plans, lifted from that day's copy of the *Times*, could be telegraphed to Russia.

The incompetence of the British government served to complicate matters; some officials quickly realized the dangers of revealing too much information, while others thought that being open with the newspapers was a good way to maintain morale and show that the government was responsive to public enthusiasm for the war. Inevitably, the government and the *Times* were soon at loggerheads. The British commander in chief, General Simpson, complained: "Our spies give us all manner of reports, while the enemy never spends a farthing for information. He gets it all for five pence from a London paper."

In addition to being the first war in which a government had to take the existence of the telegraph into account when making news public, the Crimean War was the first in which the telegraph played a strategic role. Initially, messages were sent by telegraph as far as Marseilles, and then by steamer to the Crimea, arriving as much as three weeks later. Rather than wait for a private telegraph company to step in, the British and French governments decided to extend the telegraph network to the Crimea themselves. The line was extended overland from Bucharest, the farthest extremity of the Austrian network, to Varna on the Black Sea, and a British company was then contracted to lay a 340-mile submarine cable across to the Crimean peninsula. For the first time, French and British governments could communicate directly with commanders on a distant battlefield. This was further bad news for General Simpson, who was so exasperated by trivial inquiries from his incompetent superiors in London that he is said to have complained that "the confounded telegraph has ruined everything."

For who was better placed to make strategic decisions: the commander at the scene or his distant superiors? In his history of the Crimean War, the historian A. W. Kinglake referred to the telegraph as "that new and dangerous magic" that played into the hands of meddling officials who were nowhere near the battlefield. "Our government did not abuse it," he declared, "but, exposed to swift dictation from Paris, the French had to learn what it was to carry on a war with a Louis Napoleon planted at one of the ends of the wire, and at the other, a commander like Canrobert, who did not dare to meet Palace strategy with respectful evasions, still less with plain, resolute words."

The telegraph was to cause further complications when it was used to send reports to London from the front revealing the chaotic nature of the campaign. The war was very badly organized, and although public sentiment in Britain was in favor of military action, there was widespread exasperation at the government's mismanagement, spelled out in dispatches from the front line by the *Times*'s reporter William Howard Russell. He exposed stories of soldiers being sent to the front wrongly or inadequately equipped, and highlighted the lack of proper medical support (which led to a public appeal that funded Florence Nightingale's mercy mission). It was perhaps hardly surprising that the *Times* was not allowed to use the Black Sea cable to send back its stories. Instead, reports were sent by steamer to Varna or Constantinople and then by wire to London.

The telegraph had annihilated the distance between the soldiers at the front and the readers back home, and between the government and its generals. Rather less conveniently, it had also annihilated the distance between the enemy capitals. Suddenly, the world had shrunk—

something that diplomats found particularly hard to swallow.

T raditionally, diplomats prefer slow, measured responses to events, but the telegraph encouraged instant reaction—"and I do not know that with our business it is very desirable that it should be so," warned Edmond Hammond, a British diplomat at the time of the Crimean War. He feared that diplomats would end up responding to "off-hand points which had much better be considered." Charles Mazade, a French historian, even went so far as to suggest that the Franco-Prussian War of 1870–71 was a direct result of diplomats reacting too hastily to telegraphic dispatches. But they had no choice; once the newspapers got hold of news, they would demand a statement from the government, which would then find its way into the hands of foreign governments via the media, circumventing conventional diplomatic channels.

There was only one thing for it: Diplomats would have to embrace the telegraph. So they did, albeit slowly. Until 1859, the British Foreign Office was just another customer at the telegraph office, and sent messages only during business hours; but by 1870, there were permanent lines installed at the Foreign and Colonial offices. Some officials were so keen on using the telegraph that they even had lines installed at their London homes and country houses, so that they could stay in touch with goings-on around the world. The effect was to centralize power in London; and for officials in distant countries who found their independence from central government undermined by the telegraph, the new technology was a curse. Sir Horace Rumbold, the British ambassador in Vienna, lamented "the telegraphic demoralization of those who formerly had to act for themselves."

But despite its adoption by diplomats, the telegraph was used to order troops into battle just as often as it was used to defuse a crisis. It was widely deployed during the American Civil War, with soldiers on each side stringing up a total of 15,000 miles of telegraph wire as they advanced, and engaging in much skulduggery with tapped wires and secret codes. Similarly, the telegraph proved its value as a military tool in Europe, where it was used by the Prussians to coordinate a pincer movement that led directly to their victory over the French at Könniggrätz.

Nevertheless, many people were still fervent believers in the peacemaking potential of the telegraph. In 1894, Sir John Pender, chairman of the company that had previously been the Gutta Percha Company and is known today as Cable & Wireless, suggested that telegraphy had "prevented diplomatic ruptures and consequent war, and been instrumental in promoting peace and happiness. . . . no time was allowed for the growth of bad feeling or the nursing of a grievance. The cable nipped the evil of misunderstanding leading to war in the bud."

Well, sort of. But sometimes misunderstanding was deliberate. In 1898, the Fashoda Incident, a standoff between the British and French armies in Sudan, illustrated the new power of information—and disinformation. French forces led by Major Jean-Baptiste Marchand were crossing Africa with the intention of laying claim to land from the Atlantic to the Red Sea, while a rival British expedition led by Lord Kitchener was hoping to establish control over the whole of East Africa, from Cairo to the southern Cape. Inevitably, the paths of the two armies crossed—at the Sudanese village of Fashoda. Rather than risk starting a war between two major powers, Kitchener and Marchand decided that the whole business was best left to the governments of France and Britain to sort out through diplomatic channels.

But Kitchener had a crucial advantage over Marchand: access to the British-controlled Egyptian telegraph network. He was able to send to London an immediate report on the situation, which traveled via the Egyptian railway telegraph network and then by submarine cable. He then followed up with a more detailed report, in which he suggested that Marchand's forces, which were in fact comparable in strength to his own, were demoralized, anxious, and in danger of running out of water—none of which was strictly true. But Marchand's only means of communication with his superiors in Paris was to send a messenger overland to the Atlantic coast and then on by sea—a process that would have taken nine months. As a result, the first the French government heard of the matter was when the British ambassador in Paris read Kitchener's report to the French foreign minister. Anxious to hear Marchand's side of the story, the French asked for permission to communicate with Fashoda via the British-controlled telegraph lines. The British refused but offered a compromise: If Marchand sent a messenger to Cairo, he could send messages from there. In the month it took for Marchand's representative to reach Cairo and file his report, the French had only Kitchener's version of events to go on, and took the decision to back down. The telegraph had, arguably, prevented bloodshed, if only through the use of misinformation.

OPTIMISM ABOUT THE peacemaking potential of the telegraph was still widespread at the close of the century, even though there was no evidence that it had made any real difference one way or the other. "If the peoples have been brought more in touch with each other, so also have their rulers and statesmen," wrote the British electrician and telegraph expert Charles Bright in his book, *Submarine Telegraphs*, published in 1898. "An entirely new and much-improved method of conducting diplomatic relations between one country and another has come into use with the telegraph wire and cable. The facility and rapidity with which one government is now enabled to know the 'mind'—or, at any rate, the professed mind—of another, has often been the means of averting diplomatic ruptures and consequent wars during the last few decades. At first sight, the contrary result might have been anticipated; but, on the whole, experience distinctly pronounces in favour of the pacific effects of telegraphy."

Further optimism arose from the feeling of shared experience felt by newspaper readers around the world as they followed unfolding events. One such example was the slow and lingering death of President James Garfield in 1881, two months after being shot and wounded.

In an article published in 1881, *Scientific American* assessed the "moral influence of the telegraph," which had enabled a global community to receive regular updates about his condition. Citing this as "a signal demonstration of the kinship of humanity," the article explained how "the touch of the telegraph key welded human sympathy and made possible its manifestation in a common universal, simultaneous heart throb. We have just seen the civilized world gathered as one family around a common sick bed, hope and fear alternately fluctuating in unison the world over as hopeful or alarming bulletins passed with electric pulsations over the continents and under the seas." It was, the magazine declared, "a spectacle unparalleled in history; a spectacle impossible on so grand a scale before, and indicative of a day when science shall have so blended, interwoven and unified human thoughts and interests that the feeling of universal kinship shall be, not a spasmodic outburst of occasional emotion, but constant and controlling, the usual, everyday, abiding feeling of all men toward all men."

This sort of hyperbole shows just how easy it was to assume that world peace would inevitably follow from shared experience. As one writer put it in 1878, the telegraph "gave races of men in various far-separated climes a sense of unity. In a very remarkable degree the telegraph confederated human sympathies and elevated the conception of human brotherhood. By it the peoples of the world were made to stand closer together." The rapid distribution of news was thought to promote universal peace, truthfulness, and mutual understanding. In order to understand your fellow men, you really couldn't have too much news.

Or could you? Not everyone wanted to know what was going on in far-flung countries. The precedence given to what it saw as irrelevant foreign news over important local stories even led the *Alpena Echo*, a small newspaper in Michigan, to cut off its daily telegraph service in protest. According to a contemporary account, this was because "it could not tell why the telegraph company caused it to be sent a full account of a flood in Shanghai, a massacre in Calcutta, a sailor fight in Bombay, hard frosts in Siberia, a missionary banquet in Madagascar, the price of kangaroo leather from Borneo, and a lot of nice cheerful news from the Archipelagoes—and not a line about the Muskegon fire." The seeds had been sown for a new problem: information overload.

10.

information overload

At its very birth, the telegraph system became the handmaiden of commerce.

—*NATIONAL TELEGRAPH REVIEW AND OPERATOR'S COMPANION*, 1853

Is more information always a good thing? Certainly in business, the more you know the better, and the more information you have access to, the greater your advantage over your competitors. Knowledge—about distant markets, the rise and fall of foreign empires, the failure of a crop—is, quite literally, worth money. But those in business, traditionally thirsty for the latest news, got more than they bargained for from the telegraph.

Messages between New York and Chicago, which had previously taken a month to arrive, could be delivered almost instantly; national and global markets were galvanized by the increasing flow of information. Any business that wanted to stay competitive had no choice but to embrace the new technology. The result was an irreversible acceleration in the pace of business life, which has continued to this day. And it led to a new and unexpected problem, as W. E. Dodge, a New York businessman, explained in a speech in 1868. "If the army and navy, diplomacy, science, literature and the press can claim special interest in the telegraph, surely the merchant must have as deep an interest," he said, "but I am not prepared to say that it has proved to be an unmixed blessing."

Before the telegraph, Dodge explained, New York merchants dealing in international commerce received updates from their foreign associates once or twice a month, though the information obtained in this way was usually several weeks old by the time it arrived. Those involved in national trade would be visited by their country customers twice a year on their semiannual visits to the city, and spent the summer and winter resting, looking over accounts and making plans for the future. "Comparatively, they had an easy time," said Dodge.

"But now all this is changed, and there are doubts whether the telegraph has been so good a friend to the merchant as many have supposed. Now, reports of the principal markets of the world are published every day, and our customers are continually posted by telegram. Instead of making a few large shipments in a year, the merchant must keep up constant action, multiplying his business over and over again. He has to keep up constant intercourse with distant correspondents, knows in a few weeks the result of shipments which a few years ago would not have been known for months, orders the proceeds invested in commodities, the value of which is well understood, and which are again sold before their arrival. He is thus kept in continual excitement, without time for quiet and rest.

"The merchant goes home after a day of hard work and excitement to a late dinner, trying amid the family circle to forget business, when he is interrupted by a telegram from London, directing, perhaps, the purchase in San Francisco of 20,000 barrels of flour, and the poor man must dispatch his dinner as hurriedly as possible in order to send off his message to California. The businessman of the present day must be continually on the jump, the slow express train will not answer his purpose, and the poor merchant has no other way in which to work to secure a living for his family. He *must* use the telegraph."

The information supplied by the telegraph was like a drug to businessmen, who swiftly became addicted. In combination with the railways, which could move goods quickly from one place to another, the rapid supply of information dramatically changed the way business was done.

Suddenly, the price of goods and the speed with which

they could be delivered became more important than their geographic location. Tradesmen could have several potential suppliers or markets at their disposal and were able to widen their horizons and deal directly with people whom it would have taken days to reach by mail. Direct transactions between producers and customers were made possible without having to go through middlemen; retailers, farmers, and manufacturers found that by bypassing intermediaries they could offer more competitive prices and save on commissions paid to wholesalers. Suppliers could keep smaller inventories, since there was less need to guard against uncertainties, and stock could be ordered and replenished quickly. Telegraphy and commerce thrived in a virtuous circle. "The telegraph is used by commercial men to almost as great an extent as the mail," remarked the superintendent of a telegraph line from Wall Street to Boston in 1851.

Those places beyond the reach of the network in the early days were acutely aware of their disadvantage. "The telegraph has become one of the essential means of commercial transactions," declared the *St. Louis Republican* in 1847. "Commerce, wherever lines exist, is carried on by means of it, and it is impossible, in the nature of things, that St. Louis merchants and businessmen can compete with those of other cities if they are without it. Steam is one means of commerce; the telegraph now is another, and a man may as well attempt to carry on successful trade by means of the old flatboat and keel against a steamboat, as to transact business by the use of the mails against the telegraph."

The same year, the business journalist J. D. B. De Bow noted in *The Commercial Review* "the almost incredible instances of the facilities for dispatch in business by telegraph. Every day affords instances of the advantages which our business men derive from the use of the telegraph. Operations are made in one day with its aid, by repeated communications, which could not be done in from two to four weeks by mail—enabling them to make purchases and sales which otherwise would be of no benefit to them, in consequence of the length of time consumed in negotiations."

The impact of the telegraph on commerce was greatest in the United States where a vast telegraph and rail network soon spanned the continent. "In a country in which business is spread over a vast area, and thousands of miles interpose between one commercial emporium and another, the telegraph answers to a use the complexion of which is unique," declared De Bow.

In Europe, the telegraph was seen as a public utility, and the telegraph industry wanted to maintain a balance between business and public use of the telegraph. As a result, social use of the network was more widespread than in the United States. One writer, Gardiner Hubbard, described the American telegraph system as "peculiarly a business system; eighty per cent of the messages are on business matters. . . . the managers of the telegraph know that their business customers want the quickest and best service, and care more for dispatch than low tariffs. Thus the great difference between the telegraph systems of Europe and America is that [in Europe], the telegraph is used principally for social correspondence, here by businessmen for business purposes."

The telegraph was nonetheless embraced by business in Europe. In Britain, for example, fishermen and fish traders used it to notify markets of catches and to determine market prices—something that was particularly important given the perishable nature of the goods. In Aberdeen, fish merchants were able to receive orders by telegraph while they attended sales, thanks to a pneumatic tube system that linked the fish market to the main post office. Similarly, different towns that dealt with the same commodities—such as Glasgow and Middlesborough, both of which were involved in the iron trade—became closely knit by the telegraph. The stock exchanges of principal towns were connected with the Stock Exchange in London, which was in turn linked to other exchanges around Europe and the rest of the world.

The telegraph made world produce markets a possibility; it was used to send cotton and corn prices between Liverpool, New York, and Chicago. Metal markets, ship brokering, and insurance became global businesses.

Business and telegraphy were inextricably linked. As one writer put it in 1878: "All over the earth, in every clime, throughout the territories of every civilized nation, wherever human language is known, or commerce has marts, or the smelting furnaces flash out their ever-burning fires, or the groan of giant engines work out the products of human skill or tell the story of human industry, the electric wires which web the world in a network of throbbing life utter their voices in all their varied tongues."

THE more industry and commerce came to rely upon the telegraph, the more profitable the telegraph industry itself became. Indeed, the extent of their interdependence was such that in 1870, William Orton, president of the Western Union company, which by that time had a near monopoly of the telegraph industry in the United States, told a congressional committee that the level of telegraphic traffic was as good a means as any of measuring economic activity.

"The fact is, the telegraph lives upon commerce," he said. "It is the nervous system of the commercial system. If you will sit down with me at my office for twenty minutes, I will show you what the condition of business is at any given time in any locality in the United States. This last year the grain business in the West has been very dull; as a consequence, the receipts from telegrams from that section have fallen off twenty-five per cent. Business in the South has been gaining a little, month by month, for the last year or so; and now the telegraphic receipts from that quarter give stronger indications of returning prosperity than at any previous time since the war."

Orton's statement also reflected the extent to which his company dominated the telegraph industry. Run as a franchise operation, much like a fast-food restaurant chain today, Western Union indirectly employed thousands of operators who worked for the railroad companies that were its franchisees, leading to widespread concern that too much power was concentrated in the hands of one company. By 1880, Western Union handled 80 percent of the country's message traffic and was making a huge profit.

(Not surprisingly, the company regarded its near monopoly as a good thing. Far from encouraging progress, Orton claimed, competition between rival companies had actively hindered it, resulting in a lack of "unity and despatch in conducting the telegraph business. . . . the public failed to secure everywhere the benefits of direct and reliable communication. Telegraph correspondence was not only burdened with several tariffs, but with unnecessary delays for copying and retransmission at the termini of each local line. Another serious evil which the system had to contend with was the existence of competing lines upon the more important routes. The effect is to augment the expenses without increasing the business." Western Union insisted that its monopoly was in everyone's interests, even if it was unpopular, because it would encourage standardization. "Notwithstanding the clamor in regard to telegraphic monopoly," the company's in-house journal declared in 1871, "it is the result of an inevitable law that business shall be mainly conducted under one great organization.")

In Europe, the telegraphs had in most countries been government controlled from the start, and Britain's private telegraph companies were taken into public control and absorbed by the Post Office in 1869. Having a single organization controlling a whole country's network did, admittedly, make a lot of sense; in Britain, for example, it meant that a centralized "nickname" system could be introduced. Under this scheme, companies and individuals could reserve a special word as their "telegraphic address" to make life easier for anyone who wanted to send them a telegram. Telegraphic addresses were easier to remember than full postal addresses, and after 1885 the pricing scheme was changed so that it cost more to send a message to someone with a longer address. Telegraphic addresses were assigned on a first-come, first-served basis, and a book in the main telegraph office of each town listed them alphabetically and gave the actual postal delivery address in each case. More than 35,000 telegraphic addresses had been registered at the Post Office by 1889, generating a hefty income, since an annual charge was payable for each one.

The telegraphic address was just one example of how businesses were prepared to pay extra for the latest telegraphic innovations. Private leased lines, which ran from telegraph exchanges to the post rooms of large offices and government buildings to speed the sending and delivery of telegrams, became increasingly popular. And starting in the 1870s, large companies with several offices began to lease private lines for internal communication between different sites, since internal messages could then be sent for free, and large organizations could be centrally controlled from a head office. This led to the rise of large, hierarchical companies and financial organizations—big business as we know it today.

Another of the special premium services offered by telegraph companies eager to exploit the craving for information was the delivery of regular bulletins. Companies could subscribe to a digest of the morning papers or a summary of the most recent market prices. But for some businesses, a daily or twice-daily report of the price of a commodity was not enough; they needed a more frequent fix. The demand for more frequently updated information led to the development of stock tickers: machines that spewed out information in a continuous, merciless flow.

In TIMES OF UNCERTAINTY, investors seek refuge in gold. The vast increase in the U.S. national debt during the American Civil War and the corresponding growth in the volume of paper money meant that gold was increasingly sought after during the 1860s. Since the price of gold determined the prices of other commodities, the tiniest fluctuations in its value were of immense significance to the business community and needed to be reported quickly and accurately.

A Gold Room was established at the Stock Exchange on Wall Street specifically for gold trading, where the latest price was chalked up on a board. But such was the clamor for information, with messenger boys making regular trips from nearby offices to read the figure off the blackboard, that Dr. S. S. Laws, the presiding officer of the Gold Exchange and a part-time inventor, decided a more sophisticated system was needed. Laws had studied electricity under Joseph Henry and was quickly able to devise an electrically operated "gold indicator" consisting of rotating drums marked with figures. The indicator was mounted high up on the wall of the Gold Room and was controlled via two switches, which caused the indicated price to increase or decrease a small fraction at a time. The switches also operated a second indicator, which was visible from the street outside the Gold Exchange. As the price of gold rose and fell, the indicators kept track of its movements.

While this reduced the chaos in the Gold Room itself, local merchants who wanted to know the price still had to send messenger boys down to the Exchange to read the current price from the street indicator. Some firms employed as many as twelve or fifteen boys to make regular journeys to the Stock Exchange and report back with the latest prices, which they obtained by noisily pushing and shoving their way past each other.

Laws realized that the switches could operate more than just two indicators, and had the idea of installing indicators directly into the offices of merchants and brokers, and charging a subscription fee. Having secured the rights to transmit the price of gold in this way, he left his job at the Exchange, and by the end of 1866 his Gold Indicator Company had fifty subscribers, all of whose indicators were operated in tandem from the central control switches in the Gold Room.

Then, in 1867, a telegraph operator named E. A. Callahan devised an improved indicator, which worked on a different principle. Callahan was first struck by the idea when he got caught up in a knot of shouting messenger boys as he sought shelter from a shower in the doorway of one of the Exchange buildings. "I naturally thought that much of this noise and confusion might be dispensed with," he recounted, "and that the prices might be furnished through some system of telegraphy which would not require the employment of skilled operators." But he soon found out that Laws had beaten him to it; therefore, he altered his design so that it offered a continuous printed record of the fluctuations in prices of any number of stocks, printed on a paper tape by two wheels. One wheel marked the tape with letters, and the other with numbers, and each machine could be controlled by three

An early stock ticker.

wires from a central exchange. Because Callahan's invention made a chattering sound, it was almost immediately
christened the "ticker." Soon Callahan had signed up hundreds of subscribers throughout the financial district of
New York, and his invention was an immediate success.

But the stock ticker was both a blessing and a curse.
"The record of the chattering little machine can drive a
man suddenly to the very verge of insanity with joy or de-
spair," complained one writer, "but if there be blame for
that, it attaches to the American spirit of speculation and
not to the ingenious mechanism which reads and registers
the beating of the financial pulse." A Boston businessman
was more blunt: "The letters and figures used in the lan-

guage of the tape are very few, but they spell ruin in ninety-nine million ways," he lamented.

In 1869, THOMAS EDISON, then aged twenty-one, arrived in New York looking for work. He had nowhere to stay, but thanks to his contacts in the telegraph community, he was able to spend a few nights sleeping on the floor of the battery room of Dr. Laws's Gold Indicator Company. Having previously devised an unsuccessful stock printer of his own, Edison soon figured out how the indicators and the central control system in the Gold Room worked, and he happened to be around one day when the control system suddenly made a terrible noise and stopped working. Gold prices were no longer being sent out, and each of Dr. Laws's panic-stricken subscribers, who by this stage numbered over three hundred, sent a messenger boy down to the Exchange to see what was happening.

"Within two minutes over three hundred boys crowded the office, that hardly had room for one hundred, all yelling that such and such a broker's wire was out of order and to fix it at once. It was pandemonium," Edison later recalled. He went over to the controller and figured out what the problem was: A spring had fallen out of part of the machine and got jammed between two of its gear wheels, preventing them from turning. "As I went out to tell the man in charge what the problem was, Dr. Laws appeared on the scene. He demanded of the man the cause of the trouble, but the man was speechless. I ventured to say that I knew what the trouble was, and he said 'Fix it! Fix it! Be quick!' " Edison pulled out the spring, reset the machine, and before long everything was working again.

The following day, Edison went to see Dr. Laws and suggested a number of improvements and ways in which his design could be simplified so that the system would be less likely to break down again. Suitably impressed, Laws decided to put Edison in charge of the whole operation, on a salary of $300 a month. To Edison, who was penniless and unemployed at the time, this was an absolute fortune.

Soon afterward, Laws's company merged with Callahan's, so Edison decided to start his own company. He teamed up with Franklin Pope, another young engineer who had previously worked for Dr. Laws, and the two of them went into business installing private telegraph lines and supplying specialist telegraphic equipment to business users. They also devised a stock ticker that needed only one wire, rather than the usual three, and offered a special service to importers and exchange brokers that provided just gold and sterling prices, but at a far lower price than a full stock ticker. Eventually this company too was absorbed by Callahan's business, by now known as the Gold and Stock Telegraph Company.

Edison's resourcefulness soon came to the attention of General Marshall Lefferts, the president of the Gold and Stock Telegraph Company, who offered to fund Edison's research in return for the right to use his inventions. Both parties benefited from the arrangement: Edison was able to devote all his time to inventing, and the machines he supplied gave the company a decided commercial advantage over its competitors. Before long, Edison had devised further improvements to the stock ticker, including a cunning device that enabled tickers that got out of step to be reset from the central station, without the need to send an engineer to the broker's office. Anxious to prevent anyone else from exploiting such inventions, Lefferts decided to offer Edison a cash payment for the patent rights. He offered $40,000, a figure that was so much more than Edison had expected that he almost fainted. (Edison's inexperience with large sums of money was highlighted when he cashed the check and was presented with the entire amount in small bills by a mischievous bank clerk.)

In a very short time, Edison had gone from poverty to financial independence. He rented a large workshop and was soon employing fifty men to build stock tickers and other equipment. Such was his obsession with quality that on one occasion Edison locked his workforce in the workshop until they had finished building a large order of stock tickers, with "all the bugs taken out." His improved stock tickers were soon being used in major cities all over the United States, and on the London Stock Exchange.

Although today Edison is principally remembered for inventing the phonograph and the light bulb, it was his telegraphic background and the enhancements he made to the stock ticker that gave him the financial freedom to pursue his career as an inventor.

But, ironically, it was the improvements that he and other inventors devised that would eventually lead to the demise of the telegraph and the community that had grown up around it; for any industry founded on a particular technology faces the danger that a new invention will render it obsolete.

11.

DECLINE
AND FALL

The highway girdling the earth is found in
the telegraph wires.
   —Tribute to Samuel Morse,
      the Father of the Telegraph, 1871

On JUNE 10, 1871, a bronze statue of Samuel Morse was unveiled in Central Park amid cheering crowds, speeches, and the strains of a specially composed "Morse Telegraph March." The statue had been funded by donations sent in by telegraph operators around the world to acknowledge their gratitude to Morse, then eighty, as the Father of the Telegraph. It was a title Morse had fought hard to defend.

For despite the use of his invention around the world, Morse started off with surprisingly little to show for it. Admittedly, he made enough to buy a house in 1847—a large villa in the Italian style on the banks of the Hudson near Poughkeepsie, surrounded by two hundred acres of land, which he named Locust Grove. The following year, at the age of fifty-seven, and twenty-three years after the death of his first wife, he was married for the second time, to a woman thirty years his junior. The local telegraph company installed a telegraph line right into his study, so that, in the words of one of his friends, Morse was "like an immense spider in the center of the vast web he himself had woven. Here he could hold court with the world."

But by the 1850s, although Morse was in a comfortable position financially, he was being cheated of the spoils of his invention. As the holder of the telegraph patent rights within the United States, he was entitled to royalties from any company that used his invention; but very few of the dozens of telegraph companies that sprang up to meet the explosive demand for telegraphy honored his patent. Instead, they used apparatuses based on subtle variations of his design, devised by rival inventors who disputed his claim over the patent rights. Numerous scientists and inventors crawled out of the woodwork claiming to have built working electric telegraphs before Morse or to have contributed toward his design. A controversy ensued over who had been the original inventor—and was therefore entitled to the royalties—so that Morse soon found himself estranged even from Gale and Vail, his former associates, and embroiled in a series of lengthy and costly legal battles.

The matter finally came before the Supreme Court in 1853. The court considered every aspect of electric telegraphy, from its earliest origins to its eventual adoption, and although it was clear that Morse's invention had required previous inventions and discoveries by others, Chief Justice Roger Taney said this did not detract from Morse's achievement, because nobody else had successfully fitted the pieces of the jigsaw together in the way Morse had. He had not invented the battery, discovered electromagnetism, invented the electromagnet, or figured out the correct battery configuration for long-distance telegraphy, but he had been the first to combine them all into a practical, working telegraph. The fact that Morse had received advice from others was declared to be irrelevant. "Neither can the inquiries he made, nor the information or advice he received from men of science in the course of his researches impair his right to the character of an inventor," Taney ruled. "For no man ever made such an invention without having first obtained this information, unless it was discovered by some fortunate accident. The fact that Morse sought and obtained the necessary information and counsel from the best sources, and acted upon it, neither impairs his rights as an inventor, nor detracts from his merits."

The final judgment was unequivocal: "Writing, printing or recording at a distance . . . was never invented, perfected, or put into practical operation, till it was done by Morse." Morse's patent was upheld; he was officially

declared the sole inventor of the telegraph, and the telegraph companies finally started paying him the royalties
he deserved.

Even so, Morse received no official recognition from
the U.S. government—in marked contrast to the situation
in Europe, where he spent many years making the rounds
and collecting honors and decorations. In 1851, the Morse
apparatus had been adopted as the standard for European
telegraphy, and Britain was the only country where other
forms of telegraph (the needle telegraphs devised by
Cooke and Wheatstone) were in widespread use—and even
there, the Morse system was steadily gaining ground due
to its evident simplicity. Indeed, when chairing a banquet
in Morse's honor in London in 1856, Cooke himself was
happy to admit the superiority of the Morse system. "I was
consulted a few months ago on the subject of a telegraph
for a country in which no telegraph at present exists," he
said. "I recommended the system of Professor Morse. I
believe that system to be one of the simplest in the world,
and in that lies its permanence and certainty."

Morse had honors heaped upon him by the nations of
Europe. He was made a chevalier of the Legion of Honor
by Napoleon III; he was awarded gold medals for scientific
merit by Prussia and Austria; he had further medals bestowed upon him by Queen Isabella of Spain, the king of
Portugal, the king of Denmark, the king of Italy; and the
sultan of Turkey presented him with a diamond-encrusted
Order of Glory, the "Nishan Iftichar." He was also made

an honorary member of numerous scientific, artistic, and academic institutions, including the Academy of Industry in Paris, the Historical Institute of France, and, strangely, the Archaeological Society of Belgium.

But although the countries of Europe ceremonially recognized Morse as the inventor of the telegraph, they weren't paying him any royalties—for he had failed to obtain patents in Europe during a yearlong trip to promote his invention in 1838–39. (The one exception was in France, where Morse had in fact been granted a patent, something that had been conveniently overlooked by the state-run telegraph company, which used his invention without paying for it.) Morse pointed out this incongruity to the U.S. ambassador in Paris, who took up his case, and in 1858 Morse was awarded the sum of 400,000 French francs (equivalent to about $80,000 at the time) by the governments of France, Austria, Belgium, the Netherlands, Piedmont, Russia, Sweden, Tuscany, and Turkey, each of which contributed a share according to the number of Morse instruments in use in each country or region.

In the face of such official recognition, many Americans, particularly those in the telegraphic profession, felt Morse had been slighted by his native country. Robert B. Hoover, the manager of a Western Union telegraph office, proposed that the nation's telegraphers should erect a statue in Morse's honor. The project was launched in the pages of the *Journal of the Telegraph* on April 1, 1870, and quickly won the backing of William Orton, president of the Western Union. Donations were soon pouring in from all over the country, and the enthusiasm for the scheme was such that telegraphers in other countries around the world sent contributions as well.

The following year, on the evening of the statue's unveiling, a huge banquet was held at the Academy of Music in New York in Morse's honor, followed by numerous adulatory speeches. The telegraph and its inventor were praised for uniting the peoples of the world, promoting world peace, and revolutionizing commerce. The telegraph was said to have "widened the range of human thought"; it was credited with improving the standard of journalism and literature; it was described as "the greatest instrument of power over earth which the ages of human history have revealed." As well as the speeches, there were quotations from the Bible and, inevitably, more ghastly telegraph poetry. Morse, an old man whose flowing white beard lent him a distinct resemblance to Father Christmas, was heralded as a "true genius," as "America's greatest inventor," and, of course, as the Father of the Telegraph.

Finally, at 9 P.M., all the telegraph wires of the United States were connected to a single Morse key, on which Morse himself bade farewell to the community he had created. "GREETINGS AND THANKS TO THE TELEGRAPH FRATERNITY THROUGHOUT THE WORLD. GLORY TO GOD IN THE HIGHEST, ON EARTH PEACE. GOOD WILL TO MEN," ran the message, transmitted by a skilled operator, after which Morse himself sat down at the operating table to tremendous cheers, which

On June 10, 1871, Samuel Morse, hailed as the Father of the Telegraph, bids farewell to the telegraph community.

were silenced by a gesture from Orton. In total silence, Morse then tapped out his signature, "S. F. B. MORSE," and the entire audience rose to its feet in a standing ovation. When the applause and cheering finally died down, Orton said, "Thus the Father of the Telegraph bids farewell to his children."

For the rest of the evening, congratulatory messages flooded in over the telegraph network from all corners of the United States and the rest of the world: from Havana, from Hong Kong, from India, from Singapore, and from Europe. People lined up to shake Morse's hand. When the festivities ended at midnight, a magnificent—natural—auroral display is said to have appeared in the sky.

But the day of celebration and hyperbole that culminated in Morse's telecast farewell was the high-water mark of the telegraph. The triumph of the telegraph, despite the initial bewilderment and skepticism that greeted it, had demonstrated the futility of resisting the inevitable; further technological advances were shortly to have a devastating impact on the telegraph and the community that had sprung up around it.

THE FIRST SIGN OF CHANGE was the telegraph companies' growing enthusiasm for automatic telegraphy, which started to gain ground in the 1870s. Automatic telegraphs—machines that could send messages without the need for skilled operators—had been around for many years, but as the level of traffic increased on busy parts of the network, the prospect of using machines to send messages faster and more reliably than human operators could became increasingly attractive.

The earliest automatic telegraphs were relatively clumsy affairs, devised by inventors who thought Morse code was too complicated to be learned by members of the public. One of the most successful attempts to make telegraphy easy enough for anyone was Wheatstone's ABC telegraph, which he patented in 1858. It consisted of two circular dials, each with a pointer like the hand of a clock, and marked with the letters of the alphabet; the upper dial was used to indicate incoming messages, and outgoing messages were spelled out on the lower dial, which was surrounded by a set of buttons. Sending a message involved pressing the button next to each letter in turn and turning a handle until the pointer moved to that letter, at which point it was prevented from turning further. Pulses of current sent down the line cause the pointer on the upper dial at the other end to indicate the same letter, and also rang a bell, to call attention to an incoming message. The ABC telegraph, known as the "communicator," was used extensively for point-to-point communication on thousands of private lines in Britain, since it had the advantage that no operator was required. It was used by businessmen and state officials, including the commissioner of police at Scotland Yard, who sat "spider-like in a web of co-extension with the metropolis" as he monitored reports coming in from all over London. Members of the royal family also had their own private lines installed.

Another popular automatic system was devised by David Hughes, a professor of music in Kentucky. Appropriately enough, given his musical background, the Hughes printer, launched in 1855, had a pianolike keyboard with alternating white and black keys, one for each letter (the modern QWERTY keyboard was not invented until twenty years later). It worked on a similar principle to that of the ABC telegraph, but with a constantly rotating "chariot," driven by clockwork, which was stopped in its tracks whenever a key was held down at the sending station. At the same moment an electromagnet activated a hammer, printing a character on a paper tape. The Hughes printer could be operated by anyone—it simply involved pressing the letter keys in order—and it provided a printed message that anyone could read, without the need for an operator at the receiving end. Although the original design was crude and technical limitations restricted its range to short distances, it was later improved to work over long lines, and was used in Britain, France, Italy, Switzerland, Austria, and Prussia.

Although these systems were easy to use, they weren't as fast as a Morse key in the hands of an experience telegrapher. Their use was also limited by their total incompatibility with Morse equipment. But in 1858, Wheatstone patented an automatic sender that could transmit messages in Morse at very high speed from a prepunched tape. This was a direct replacement for a human telegrapher, and it was capable of up to four hundred words per minute—ten times faster than the finest human operators. At the receiving end, messages were printed out as dots and dashes by a standard Morse printer, and could then be decoded into letters and numbers in the usual way. Admittedly, messages had to be punched onto a tape by hand before sending, but this was less skilled work than operating a Morse key, and it could be done in advance; long messages could be punched by several operators in parallel, each punching a different paragraph, and then spliced together.

The Wheatstone Automatic telegraph was widely com-

Wheatstone Automatic transmitter. Messages were prepunched on a paper tape and then transmitted in Morse code at very high speed as the tape was fed through the machine.

pared with the Jacquard loom, which wove cloth into a pattern determined by holes punched in cards—indeed, it was sometimes referred to as the "electric Jacquard." Following its invention, Wheatstone further refined his design, and it entered widespread use after 1867, particularly for news transmission, for which it was particularly well suited; news needed to be sent quickly because its value rapidly diminished. The Automatic was certainly fast: One night in 1886, following the introduction of Prime Minister William Gladstone's Bill for Home Rule in Ireland, no fewer than 1.5 million words were dispatched from the central telegraph station in London by one hundred of

Wheatstone's transmitters. The Automatic also dramatically increased the amount of traffic that could be carried by a telegraph line; instead of charging by the word, messages were charged by the yard of tape.

A further boost to network capacity was provided by the invention of the duplex, a means of sending messages in both directions over a single wire simultaneously. The search for a means of making the local receiver insensitive to signals sent by the transmitter had been going on since 1853, when Wilhelm Gintl of the Austrian State Telegraph devised an unsuccessful design. It was not until 1872, by which time electrical theory had advanced dramatically, that Joseph B. Stearns of Boston was able to build and patent a working duplex. It meant that telegraph companies were suddenly able to send twice as much traffic along a single wire, merely by installing special equipment at each end, and it saved them a fortune, since it cost far less to buy a set of duplex equipment than it did to string up a new wire.

Meanwhile the French, as usual, were doing things their own way. In 1874, Jean Maurice Emile Baudot of the French telegraph administration devised a novel form of automatic telegraph that squeezed even more capacity out of a telegraph line. At each end of the wire, synchronized rotating distributor arms switched the use of a single telegraph line between four or six sets of apparatus. In conjunction with duplex equipment, this enabled a single line to carry up to twelve lines' worth of traffic. Instead of

Morse code, the apparatus used a five-unit binary code, in which each letter was represented as a series of five current pulses, each of which could be positive or negative. Operators seated at each of the Baudot terminals sent messages by holding down different chords in succession on a special pianolike keyboard with five keys. During each revolution of the distributor arm, the line was automatically switched between each of the operators in turn, and five pulses were transmitted depending on whether the keys were up (negative pulse) or down (positive pulse). The distributor arm typically rotated two or three times a second, and each operator had the use of the line for a fraction of a second during each revolution, so timing was critical; a clicking noise made at the start of each revolution helped operators with their split-second timing. At the receiving end, an ingenious electromechanical device turned the stream of pulses into a message printed in roman type on a paper tape.

The Baudot telegraph was capable of up to 30 words per minute on each line; with twelve sets of apparatus at each end, the effective capacity of a single wire was 360 words per minute. However, operating a Baudot terminal was very stressful, due to the need for accurate timing, so on average only around two-thirds of this speed was attained. But since there was no need for a receiving operator, the Baudot halved the number of skilled operators required to send a message.

The same year, Edison, who had been beaten by

Stearns in the race to invent the first duplex system, invented the quadruplex, which, as its name suggests, enabled a single wire to carry four streams of traffic, essentially by superimposing two duplex circuits. The trick was to find a way to send two messages at once in the same direction; this was achieved by using one set of apparatus that was sensitive to changes in the direction of electrical current, and another sensitive to large steps in the magnitude of the current. Like the duplex, the quadruplex was rapidly adopted for obvious reasons. It was far cheaper to add a gizmo to each end of a wire than it was to string up three new wires. In fact, the "quad" was said to have saved the Western Union company $500,000 a year in the construction of new lines.

The combination of these new technologies enabled telegraph companies to save money on construction and skilled labor; reducing operating costs while making maximum use of network capacity was the name of the game. One study in 1883 pointed out that automatic telegraphy from prepunched tapes offered dramatic reductions in labor costs: Unskilled workers could operate the machines for just a quarter of the salary of a highly skilled Morse key operator.

Thanks to the relentless pace of technological change, telegraphy was changing from a high-skill to a low-skill occupation; from a carefully learned craft to something anyone could pick up. As the emphasis switched from skilled operators to the latest high-tech equipment, the tone of the telegraphic journals changed; the humorous stories and telegraph poetry were replaced by circuit diagrams and lengthy explanations aimed at technical and managerial readers, rather than the lowly minions who merely operated the machines. The growing use of automatic machinery was undermining the telegraphic community; and another new invention was to deal it a deadly blow.

THE INVENTION OF the duplex and the quadruplex had shown that with the right sort of electrical trickery, a single telegraph wire could be made to carry the traffic of two and then four wires. Could a wire be made to carry even more traffic? Anyone who could find a way to improve the quadruplex would have a ready market for his invention, since it would save the telegraph companies huge amounts of money. Not surprisingly, many inventors devoted much time and effort in the search for new ways to squeeze ever more telegraph traffic into a single wire.

One approach, which was being pursued by several inventors, was known as the "harmonic" telegraph. The human ear can distinguish notes of different pitches, and if each of those notes is playing a separate rhythm, anyone of a sufficiently musical turn of mind can "tune out" all but one of the notes—just as it is possible to separate the voice of someone at a party from the hubbub of the sur-

Elisha Gray, the inventor whose work on a harmonic telegraph contributed to the invention of the telephone.

rounding crowd. The idea of the harmonic telegraph was
to use a series of reeds vibrating at different frequencies.
Electrical signals produced by the reeds would be combined, sent down a telegraph wire, and then separated out
again at the other end using an identical set of reeds, each
of which would respond only to the signals generated by
its counterpart. Morse telegraphy would then be possible
by stopping and starting the vibration of each reed to make
dots and dashes.

Elisha Gray, one of those working on a harmonic telegraph, produced a design that he believed would be capable of carrying sixteen messages along a single wire. But
when he tested his design, he found that in practice only
six separate signals could be sent reliably. Nevertheless,
Gray was confident that he would eventually be able to improve his apparatus.

Another inventor working on a harmonic telegraph

was Alexander Graham Bell. He was testing his equipment on June 2, 1875, when one of the reeds got stuck and his assistant, Thomas Watson, plucked it much harder than usual in order to free it. Bell, listening at the other end of the wire, heard the unmistakable twang of the reed—a far more complex sound than the pure musical tones his apparatus had been designed to transmit. Bell realized that, with a few modifications, his apparatus might be capable of far more than mere telegraphy. It looked as though he had stumbled upon a way of transmitting any sound—including the human voice—along a wire from one place to another.

Bell worked for several months to build a working prototype. On February 14, 1876, when it became clear that Gray was pursuing the same goal, Bell filed for a patent, even though he had yet to successfully transmit speech. He was granted the patent on March 3, and made the vital breakthrough a week later, when he succeeded in transmitting intelligible speech for the first time. After several months of further refinement, his new invention—the telephone—was ready for the world.

Initially, the telephone was seen merely as a "speaking telegraph"—an improvement of an existing technology, rather than something altogether different. Even Bell, whose 1876 patent was entitled "Improvements in Telegraphy," referred to his invention as a form of telegraph in a letter to potential British investors. "All other telegraphic machines," he wrote, "produce signals which require to be

Alexander Graham Bell,
inventor of the telephone.

translated by experts, and such instruments are therefore extremely limited in their application. But the telephone actually speaks." Gray's lawyers advised him that the telephone was merely an unimportant by-product in the far more important race to build a harmonic telegraph, so, initially at least, he did not contest Bell's right to the telephone patent. It was a decision he soon came to regret.

The advantages of the telephone over all forms of telegraph were clear, and they were spelled out in the first advertisement for telephone service, issued by the newly formed Bell Telephone Company, in May 1877: "No skilled operator is required; direct conversation may be had by speech without the intervention of a third person. The communication is much more rapid, the average number of words being transmitted by Morse Sounder being from fifteen to twenty per minute, by Telephone from one to two hundred. No expense is required either for its operation, maintenance, or repair. It needs no battery and has no complicated machinery. It is unsurpassed for economy and simplicity."

The telephone was an instant success. By the end of June 1877, there were 230 telephones in use; a month later, the figure was 750; a month later still, there were 1,300. By 1880, there were 30,000 telephones in use around the world.

Meanwhile, a host of new electrical innovations, such as the use of electrical sparks to light gas lamps in large buildings, were being devised, though they were initially regarded, like the telephone, as mere spin-offs of telegraphy. But with Edison's invention of the incandescent lightbulb in 1879, and the use of electricity for everything from lighting to powering electric trams and elevators, it became clear that telegraphy was merely one of many applications of electricity—and a comparatively old-fashioned one at that. Abandoning his telegraphic roots, Edison turned his attention to electrical matters, devising ever more efficient forms of generator to supply household electricity and inventing the electricity meter to monitor its use.

As enthusiasm for all things electrical blossomed in

the 1880s and the telephone continued its rapid growth, the telegraph was no longer at the cutting edge of technology. "So much have times altered in the last fifty years that the electric telegraph itself is threatened in its turn with serious rivalry at the hands of a youthful and vigourous competitor. A great future is doubtless in store for the telephone," declared *Chambers Journal* in 1885.

By this time, many telegraphers were complaining that they had been reduced to mere machines, while others decried the declining quality of those entering the profession. "The character of the business has wholly changed," lamented the *Journal of the Telegraph*. "It cannot now subserve public interests or its own healthful development without the precision and uniformity of mechanism."

But the changing fortunes of telegraphy were perhaps most vividly illustrated in the way the telegraphic journals, which had covered the rise of the new electrical and telephonic technologies with keen interest, chose to rename themselves: the *Telegraphers' Advocate* became the *Electric Age*, the *Operator* renamed itself *Electrical World*, and the *Telegraphic Journal* became the *Electrical Review*. Undermined by the relentless advance of technology, the telegraphic community, along with its customs and subculture, began to wither and decline.

12.

THE LEGACY
OF THE TELEGRAPH

What now my old telegraph,
At the top of your old tower,
As somber as an epitaph,
And as still as a boulder?
   — from "Le Vieux Télégraphe," a poem by
   Gustave Nadaud, translated by the author

Morse never saw the birth of the inventions that would overshadow the telegraph. He agreed to unveil a statue of Benjamin Franklin in Printing House Square in New York, but exposure to the bitterly cold weather on the day of the ceremony weakened him considerably. As he lay on his sickbed a few weeks later, his doctor tapped his chest and said, "This is the way we doctors telegraph, Professor." Morse smiled and replied, "Very good, very good." They were his last words. He died in New York on April 2, 1872, at the age of eighty-one, and was buried in the Greenwood Cemetery. Shortly before his death, his estate had been valued at half a million dollars—a respectable sum, though less than the fortunes amassed by the entrepreneurs who built empires on the back of his invention. But it was more than enough for Morse, who had given freely to charity and endowed a lectureship on "the relation of the Bible to the Sciences."

Arguably, the tradition of the gentleman amateur scientist died with him. The telegraph had originated with Morse and Cooke, both of whom combined a sense of curiosity and invention with the single-mindedness needed to get it off the ground; it had then entered an era of consolidation, during which scientists like Thomson and Wheatstone provided its theoretical underpinnings; and it had ended up the province of the usual businessmen who take over whenever an industry becomes sufficiently stable, profitable, and predictable. (Edison might appear to have much in common with Morse and Cooke on the surface, but he was no amateur; he could never have devised the quadruplex without a keen understanding of electrical theory, something that both Morse and Cooke lacked.)

Wheatstone died in 1875, having amassed many honors and a respectable fortune from the sale of his various telegraphic patents. Like Morse, he was made a chevalier of the Legion of Honor, and he was knighted in 1868 following the success of the Atlantic cable. By the time of his death, he had received enough medals to fill a box a cubic foot in capacity—and he was still not getting on with Cooke. Wheatstone refused the offer of the Albert Medal from the Royal Society of Arts, because it was also offered to Cooke, and Wheatstone resented the implication of equality. He continued to work as a scientist, with particular interest in optics, acoustics, and electricity, and he died a rich and well-respected man. Alongside his innovations in the field of telegraphy, he invented the stereoscope and the concertina, though today his name is known to students via the Wheatstone bridge, a method of determining electrical resistance that, somewhat characteristically, he did not actually invent but helped to popularize.

Cooke, on the other hand, failed to distinguish himself after his promising start; indeed, it's not hard to see why Wheatstone so resented being compared to him. He worked as an official of the Electrical Telegraph Company from its establishment in 1845 until it was taken into state control by the British government in 1869, when he was knighted. But he was soon in financial difficulties. He bought a quarry and poured the money he obtained from the sale of his share of the telegraph patent into a handful of abortive new inventions, including stone- and slate-cutting machines, and a design for a rope-hauled railway with remote-controlled doors that he unsuccessfully tried to have adopted on the underground railway in London. The prime minister, William Gladstone, was alerted to his plight and granted him a £100 annual state pension, the maximum possible amount. But it was not enough to keep

An early telephone exchange.

Cooke out of debt. His rivalry with Wheatstone continued until Wheatstone's death; Cooke attended his funeral and was, curiously, far more accurate in his recollection of Wheatstone's role in the invention of the telegraph thereafter. He died in 1879, having squandered his fortune.

By the late 1880s, the telephone was booming. In 1886, ten years after its invention, there were over a quarter of a million telephones in use worldwide. Early technological hurdles such as low sound quality, long-distance calling, and the design of efficient manual and automatic telephone exchanges were rapidly overcome by Edison, Hughes, Watson, and others, and by the turn of the century there were nearly 2 million phones in use. (Bell did little to improve his invention; once its success was assured, he turned his attention to aviation instead.)

When Queen Victoria's reign ended in 1901, the telegraph's greatest days were behind it. There was a telephone in one in ten homes in the United States, and it was being swiftly adopted all over the country. In 1903, the English inventor Donald Murray combined the best features of the Wheatstone and Baudot automatic telegraph systems into a single machine, which, with the addition of a typewriter keyboard, soon evolved into the teleprinter. Like the telephone, it could be operated by anyone.

The heyday of the telegrapher as a highly paid, highly skilled information worker was over; telegraphers' brief tenure as members of an elite community with mastery over a miraculous, cutting-edge technology had come to an end. As the twentieth century dawned, the telegraph's inventors had died, its community had crumbled, and its golden age had ended.

ALTHOUGH IT HAS now faded from view, the telegraph lives on within the communications technologies that have subsequently built upon its foundations: the telephone, the fax machine, and, more recently, the Internet. And, ironically, it is the Internet—despite being regarded as a quintessentially modern means of communication—that has the most in common with its telegraphic ancestor.

Like the telegraph network, the Internet allows people to communicate across great distances using interconnected networks. (Indeed, the generic term *internet* simply means a group of interconnected networks.) Common rules and protocols enable any sort of computer to exchange messages with any other—just as messages could easily be passed from one kind of telegraph apparatus (a Morse printer, say) to another (a pneumatic tube). The journey of an e-mail message, as it hops from mail server to mail server toward its destination, mirrors the passage of a telegram from one telegraph office to the next.

There are even echoes of the earliest, most primitive telegraphs—such as the optical system invented by Chappe—in today's modems and network hardware. Every time two computers exchange an eight-digit binary number, or byte, they are going through the same motions as an eight-panel shutter telegraph would have done two hundred years ago. Instead of using a codebook to relate each combination to a different word, today's computers use another agreed-upon protocol to transmit individual letters. This scheme, called ASCII (for American Standard Code for Information Interchange), says, for example, that a capital "A" should be represented by the pattern 01000001; but in essence the principles are unchanged since the late eighteenth century. Similarly, Chappe's system had special codes to increase or reduce the rate of transmission, or to request that garbled information be sent again—all of which are features of modems today. The protocols used by modems are decided on by the ITU, the organization founded in 1865 to regulate international telegraphy. The initials now stand for International Telecommunication Union, rather than International Telegraph Union.

More striking still are the parallels between the social impact of the telegraph and that of the Internet. Public reaction to the new technologies was, in both cases, a confused mixture of hype and skepticism. Just as many Victorians believed the telegraph would eliminate misunderstanding between nations and usher in a new era of world peace, an avalanche of media coverage has lauded the Internet as a powerful new medium that will transform and improve our lives.

Some of these claims sound oddly familiar. In his 1997 book *What Will Be: How the New World of Information Will Change Our Lives*, Michael Dertouzos of the Laboratory for Computer Science at the Massachusetts Institute of Technology wrote of the prospect of "computer-aided peace" made possible by digital networks like the Internet. "A common bond reached through electronic proximity may help stave off future flareups of ethnic hatred and national breakups," he suggested. In a conference speech in November 1997, Nicholas Negroponte, head of the MIT Media Laboratory, explicitly declared that the Internet would break down national borders and lead to world peace. In the future, he claimed, children "are not going to know what nationalism is."

The similarities do not end there. Scam artists found crooked ways to make money by manipulating the transmission of stock prices and the results of horse races using the telegraph; their twentieth-century counterparts have used the Internet to set up fake "shop fronts" purporting to be legitimate providers of financial services, before disappearing with the money handed over by would-be investors; hackers have broken into improperly secured computers and made off with lists of credit card numbers.

People who were worried about inadequate security on the telegraph network, and now on the Internet, turned to the same solution: secret codes. Today software to compress files and encrypt messages before sending them across the Internet is as widely used as the commercial codes that flourished on the telegraph network. And just as the ITU placed restrictions on the use of telegraphic ciphers, many governments today are trying to do the same with computer cryptography, by imposing limits on the complexity of the encryption available to Internet users. (The ITU, it should be noted, proved unable to enforce its rules restricting the types of code words that could be used in telegrams, and eventually abandoned them.)

On a simpler level, both the telegraph and the Internet have given rise to their own jargon and abbreviations. Rather than plugs, boomers, or bonus men, Internet users are variously known as surfers, netheads, or netizens. Personal signatures, used by both telegraphers and Internet users, are known in both cases as sigs.

Another parallel is the eternal enmity between new, inexperienced users and experienced old hands. Highly skilled telegraphers in city offices would lose their temper when forced to deal with hopelessly inept operators in remote villages; the same phenomenon was widespread on the Internet when the masses first surged on-line in the early 1990s, unaware of customs and traditions that had held sway on the Internet for years and capable of what, to experienced users, seemed unbelievable stupidity, gullibility, and impoliteness.

But while conflict and rivalry both seem to come with the on-line territory, so does romance. A general fascination with the romantic possibilities of the new technology has been a feature of both the nineteenth and twentieth centuries: On-line weddings have taken place over both the telegraph and the Internet. In 1996, Sue Helle and Lynn Bottoms were married on-line by a minister 10 miles away in Seattle, echoing the story of Philip Reade and Clara Choate, who were married by telegraph 120 years earlier by a minister 650 miles away. Both technologies have also been directly blamed for causing romantic problems. In 1996, a New Jersey man filed for divorce when he discovered that his wife had been exchanging explicit e-mail with another man, a case that was widely reported as the first example of "Internet divorce."

After a period of initial skepticism, businesses became the most enthusiastic adopters of the telegraph in the nineteenth century and the Internet in the twentieth. Businesses have always been prepared to pay for premium services like private leased lines and value-added information—provided those services can provide a competitive advantage in the marketplace. Internet sites routinely offer stock prices and news headlines, both of which were available over a hundred years ago via stock tickers and news wires. And just as the telegraph led to a direct increase in the pace and stress of business life, today the complaint of information overload, blamed on the Internet, is commonplace.

The telegraph also made possible new business practices, facilitating the rise of large companies centrally controlled from a head office. Today, the Internet once again promises to redefine the way people work, through emerging trends like teleworking (working from a distant location, with a network connection to one's office) and virtual corporations (where there is no central office, just a distributed group of employees who communicate over a network).

The similarities between the telegraph and the Internet—both in their technical underpinnings and their social impact—are striking. But the story of the telegraph contains a deeper lesson. Because of its ability to link distant peoples, the telegraph was the first technology to be seized upon as a panacea. Given its potential to change the world, the telegraph was soon being hailed as a means of solving the world's problems. It failed to do so, of course—but we have been pinning the same hope on other new technologies ever since.

In the 1890s, advocates of electricity claimed it would eliminate the drudgery of manual work and create a world of abundance and peace. In the first decade of the twentieth century, aircraft inspired similar flights of fancy: Rapid intercontinental travel would, it was claimed, eliminate international differences and misunderstandings. (One commentator suggested that the age of aviation would be an "age of peace" because aircraft would make armies obsolete, since they would be vulnerable to attack from the air.) Similarly, television was expected to improve education, reduce social isolation, and enhance democracy. Nuclear power was supposed to usher in an age of plenty where electricity would be "too cheap to meter." The optimistic claims now being made about the Internet are merely the most recent examples in a tradition of technological utopianism that goes back to the first transatlantic telegraph cables, 150 years ago.

That the telegraph was so widely seen as a panacea is perhaps understandable. The fact that we are still making the same mistake today is less so. The irony is that even though it failed to live up to the utopian claims made about it, the telegraph really did transform the world. It also redefined forever our attitudes toward new technologies. In both respects, we are still living in the new world it inaugurated.

epilogue

T HE HYPE, skepticism, and bewilderment associated with the Internet—concerns about new forms of crime, adjustments in social mores, and redefinition of business practices—mirror the hopes, fears, and misunderstandings inspired by the telegraph. Indeed, they are only to be expected. They are the direct consequences of human nature, rather than technology.

Given a new invention, there will always be some people who see only its potential to do good, while others see new opportunities to commit crime or make money. We can expect exactly the same reactions to whatever new inventions appear in the twenty-first century.

Such reactions are amplified by what might be termed chronocentricity—the egotism that one's own generation is poised on the very cusp of history. Today, we are repeatedly told that we are in the midst of a communications revolution. But the electric telegraph was, in many ways, far more disconcerting for the inhabitants of the time than today's advances are for us. If any generation has the right to claim that it bore the full bewildering, world-shrinking brunt of such a revolution, it is not us—it is our nineteenth-century forebears.

Time-traveling Victorians arriving in the late twentieth century would, no doubt, be unimpressed by the Internet. They would surely find space flight and routine intercontinental air travel far more impressive technological achievements than our much-trumpeted global communications network. Heavier-than-air flying machines were, after all, thought by the Victorians to be totally impossible. But as for the Internet—well, they had one of their own.

afterword

One of the advantages of writing an Internet book in which everything is already 150 years out of date is that the story is in less danger of being overtaken, or rendered irrelevant, by subsequent developments. *The Victorian Internet* was mostly written in 1997, and is published here in its original form. Much has changed on the Internet since then, of course; the utopianism of the late 1990s evaporated in the dotcom crash of 2000, though the spread of broadband connections and the growth of new Internet business models built around online commerce and advertising have since helped many firms to bounce back. And despite everything that has happened in the past ten years, the analogy between the Internet and the telegraph still holds.

That the Internet has continued to evolve is not surprising. But oddly enough the telegraph has been in the news, too. Having been inaugurated with Samuel Morse's first official telegram, "What hath God wrought!" in 1844, telegraph service in the United States was discontinued in 2006 with a rather more prosaic message: "Effective January 27, 2006, Western Union will discontinue all Telegram and Commercial Messaging services. We regret any inconvenience this may cause you, and we thank you for your loyal patronage." Like many people I was saddened to hear of its demise—yet I was also surprised to learn that the telegram had survived as long as it had, given the availability of so many faster, cheaper, and more convenient forms of electronic messaging.

Ten years ago electronic messaging would have meant e-mail in particular, but another striking development of the past decade has been the curious rebirth of the telegram in the form of text messages sent between mobile phones. Initially adopted by European teenagers as a cheap alternative to expensive mobile-phone calls, text messages have become a new communications medium in their own right, all over the world. Around 1.3 trillion text messages were sent during 2006; Americans, who were relatively late adopters of the technology, sent 158 billion text messages that year. (The volume of e-mails is still higher, however; around 9 trillion non-"spam" e-mails were sent in 2006.)

Like telegrams before them, text messages force people to be brief and to the point, and they have spawned their own vocabulary of space-saving abbreviations, such as "c u l8r." This is not the only echo from the age of the telegram. A Nokia handset can be programmed to announce incoming text messages with three short beeps, two long ones, and three short ones—Morse code for "SMS" or "short message service," the technical term for text messaging. Samuel Morse would be proud. A defunct nineteenth-century technology has, in effect, been reincarnated in the twenty-first century. The telegram is dead; long live the telegram.

Yet the mobile phone is not just the heir to the telegraph's tradition. It could also prove to be the most important heir to the personal computer, as Internet-enabled mobile devices proliferate. Indeed, mobile phones could do for the Internet what the telephone did for the telegraph: make it easier to use and far more widely available.

It was the telegraph's fate to be overshadowed by several of its offspring, by the telephone in particular, which was first regarded as a minor variation of the technology (a "speaking telegraph") but ended up being far more popular. The telegraph also spawned the stock ticker, teletype machines, and early forms of a fax machine capable of sending pictures by telegraph. All of these dedicated devices used variations of the original telegraph technology for particular purposes. The same thing is now happening to the Internet, as it becomes embedded into other devices, rather than being something accessed only through a personal computer. Task-specific devices such as Internet-capable music players, games consoles, television set-top boxes, and hi-fis are now available.

But access to the Internet through mobile devices, such as phones and BlackBerry handhelds, is the biggest growth area. In rich countries, personal computers were widespread before mobile phones. In poor countries mobile phones are widespread than personal computers. As C. K. Prahalad, an Indian management guru, puts it: "Emerging markets will be wireless-centric, not PC-centric." The number of mobile phones in use is over 2.5 billion and growing fast, even in the poorest parts of the world. Mobile phones will complete the democratization of telecommunications started by the telegraph.

Tom Standage
April 2007

sources

In addition to the specific books and journals listed below, I found information on the subject in the following publications: *Electrical World* (New York), *Journal of Commerce* (New York), *Journal of the Telegraph* (New York), *Scientific American* (New York), and *The Times* (London).

Anecdotes of the Telegraph. London: David Bogue, 1849.

Babbage, Charles. *Passages from the Life of a Philosopher.* London: Longman & Co., 1864.

Blondheim, Menahem. *News over the Wires: The Telegraph and The Flow of Public Information in America, 1844–1897.* Cambridge, Mass. and London: Harvard University Press, 1994.

Bowers, Brian. *Sir Charles Wheatstone*. London: HMSO, 1975.

Briggs, Charles, and Augustus Maverick. *The Story of the Telegraph*. New York, 1858.

Clarke, Arthur C. *How the World Was One*. London: Victor Gollancz, 1992.

Clow, D. G. "Pneumatic Tube Communication Systems in London," Newcomen Transactions, pp. 97–115, 1994–95.

Coe, Lewis. *The Telegraph, a History of Morse's Invention and Its Predecessors in the United States*. London: McFarland, 1993.

Congdon, Charles. *Reminiscences of a Journalist*. Boston: James R. Osgood & Co., 1880.

Cooke, Sir William Fothergill. *The Electric Telegraph: Was It Invented by Professor Wheatstone?* (Mr. Cooke's First Pamphlet; Mr. Wheatstone's Answer; Mr. Cooke's Reply; Arbitration Papers and Drawings.) London: W. H. Smith & Son, 1857.

Corn, Joseph J., ed. *Imagining Tomorrow: History, Technology, and the American Future*. Cambridge, Mass.: MIT Press, 1986.

Dertouzos, Michael. *What Will Be. How the New World of Information Will Change Our Lives*. London: Piatkus, 1997.

Du Boff, Richard B. "Business Demand and the Development of the Telegraph in the United States, 1844–1860," *Business History Review* vol. 54, pp. 459–79, 1980.

Dyer, Frank Lewis, and Thomas Commerford Martin. *Edison: His Life and Inventions*. New York: Harper and Brothers, 1910.

Gabler, Edwin. *The American Telegrapher—A Social History*. New Brunswick: Rutgers University Press, 1988.

Headrick, Daniel R. *The Invisible Weapon*. London: Oxford University Press, 1991.

Holzmann, Gerard, and Bjorn Pehrson. *Early History of Data Networks*. Los Alamitos, Calif.: IEEE Computer Society Press, 1995.

Hubbard, G. *Cooke and Wheatstone and the Invention of the Electric Telegraph*. London: Routledge and Kegan Paul, 1965.

Kahn, David. *The Codebreakers*. New York: Macmillan, 1967.

Kieve, Jeffrey. *The Electric Telegraph: A Social and Economic History*. Newton Abbot: David and Charles, 1973.

Lebow, Irwin. *Information Highways and Byways: From the Telegraph to the 21st Century*. New York: IEEE Computer Society Press, 1995.

Marland, Edward Allen. *Early Electrical Communication*. London: Abelard-Schumann, 1964.

Marvin, Carolyn. *When Old Technologies Were New: Thinking About Electric Communication in the Late Nineteenth Century*. New York and Oxford: Oxford University Press, 1988.

Morse, Samuel F. B., and Edward Lind Morse. *Samuel F. B. Morse: His Letters and Journals, Edited and Supplemented by His Son Edward Lind Morse.* Boston: Houghton Mifflin, 1914.

Prescott, George B. *Electricity and the Electric Telegraph.* London: Spon, 1878.

Prescott, George B. *History, Theory, and Practice of the Electric Telegraph.* Boston, 1860.

Prime, Samuel Irenaeus. *The Life of Samuel F. B. Morse.* New York: Appleton, 1875.

Reid, James D. *The Telegraph in America and Morse Memorial.* New York: Derby Brothers, 1879.

Shaffner, Tal. *The Telegraph Manual.* New York: Pudney and Russell, 1859.

Shiers, George, ed. *The Electric Telegraph: An Historical Anthology.* New York: Arno Press, 1977.

Spufford, Francis, and Jenny Uglow, eds. *Cultural Babbage.* London: Faber and Faber, 1996.

Thompson, Robert L. *Wiring a Continent. The History of the Telegraph Industry in the United States, 1832–1866.* Princeton: Princeton University Press, 1947.

Turnbull, Laurence. *The Electro-Magnetic Telegraph.* Philadelphia: A. Hart, 1852.

F. H. Webb, ed. *Cooke, Sir William Fothergill: Extracts from the Private Letters of the Late Sir William Fothergill Cooke, 1836–39, Relating to the Invention and Development of the Electric Telegraph.* London: E. & F. N. Spon, 1895.

Wilson, Geoffrey. *The Old Telegraphs.* London: Chichester Phillimore, 1976.

acknowledgments

I am GRATEFUL to many people for helping to make this book possible, including the following (in order of appearance): my wife, Kirstin, Chester, Azeem Azhar, Ben Rooney and Roger Highfield at the *Daily Telegraph*, Oliver Morton, Virginia Benz and Joe Anderer, Katinka Matson and John Brockman, George Gibson, Jackie Johnson, Mary Godwin at the Cable & Wireless Archive, Ravi Mirchandani, and Georgia Cameron-Clarke.

INDEX

a note on the author

Tom Standage is business editor at the *Economist* and the author of *A History of the World in 6 Glasses*, *The Victorian Internet*, *The Turk*, and *The Neptune File*. *The Victorian Internet* was made into a documentary, "How the Victorians Wired the World." He has written about science and technology for numerous magazines and newspapers, including *Wired*, the *Guardian*, the *Daily Telegraph*, and the *New York Times*. He lives in Greenwich, England, with his wife, daughter, and son.